William Paul Gerhard

Theatre Fires and Panics:, their Causes and Prevention

William Paul Gerhard

Theatre Fires and Panics:, their Causes and Prevention

ISBN/EAN: 9783743390713

Manufactured in Europe, USA, Canada, Australia, Japa

Cover: Foto ©Lupo / pixelio.de

Manufactured and distributed by brebook publishing software (www.brebook.com)

William Paul Gerhard

Theatre Fires and Panics:, their Causes and Prevention

THEATRE FIRES AND PANICS:

THEIR

CAUSES AND PREVENTION.

BY

WILLIAM PAUL GERHARD, C.E.,
Consulting Engineer for Sanitary Works.

FIRST EDITION.
FIRST THOUSAND.

NEW YORK:
JOHN WILEY & SONS.
LONDON: CHAPMAN & HALL, LIMITED.
1896.

Copyright, 1896,
BY
WM. PAUL GERHARD.

ROBERT DRUMMOND, ELECTROTYPER AND PRINTER, NEW YORK.

PREFACE.

THIS little book is not intended to be a general treatise on "Theatre Planning and Construction." It is simply, as its title implies, a discussion of the causes and the prevention of fires and panics in theatres. A few years ago the writer prepared several essays on the subject, which are herewith reproduced substantially as originally given. The writer has added at the end of the book a list of the literature (works, pamphlets, and articles in technical journals) bearing upon the subject, which he has found of value in studying the whole question, and which he hopes will be helpful to others.

May the book prove of practical utility and of interest generally to architects, engineers, theatre managers, and fire underwriters, and fulfil its mission of securing greater *safety in theatres!*

THE AUTHOR.

NEW YORK CITY, 36 UNION SQUARE,
      July, 1896.

CONTENTS.

I. THEATRE-FIRE CATASTROPHES AND THEIR PREVENTION.

	PAGE
Introduction	1
Theatre-fire Statistics	3
List of Twelve Prominent Theatre-fire Calamities of this Century	12
Causes of Theatre Fires	20
Causes of Panics	25
Dangers to Human Life	27
The Prevention of Theatre-fire Catastrophes	29
(*a*) Measures to prevent outbreak of fire	29
(*b*) Measures to localize fires	29
(*c*) Measures to insure the safety of the spectators and of the stage people	30
(*d*) Measures to put out fires	30
Theatre Inspections	40
Theatre Management	44
Conclusion	45

II. THE ESSENTIAL CONDITIONS OF SAFETY IN THEATRES.

Introduction	47
Causes of Theatre Fires	49
Statistics of Theatre Fires	51
General Considerations	52
Location and Site	57
Plan of Building	59
Construction of Building	65
Staircases, Entrances and Exits, and Fire-escapes	69

CONTENTS.

	PAGE
Aisles and Chairs	73
Fire-proof Curtain	75
Stage Ventilator	77
Fire-proof Treatment of Stage Scenery	79
Stage Construction and Stage Machinery	80
Heating	83
Lighting	85
Lightning-rods	89
Ventilation and Sanitation	90
Fire-service and Fire-extinguishing Appliances	92
Life-saving Appliances	97
Fire-alarms	98
Questions of Management	99
Periodical Inspection	103

III. THE WATER-SERVICE AND FIRE PROTECTION OF THEATRES.

Introduction... 105

GENERAL REQUIREMENTS OF SAFETY IN THEATRES.

1. As regards the Location................................. 105
2. As regards the Plan.................................... 106
3. As regards the Construction........................... 106
4. As regards the Interior Arrangement and Equipment....... 107
5. As regards the Management............................. 109
6. As regards Inspection................................. 111

FIRE-EXTINGUISHING APPLIANCES.

1. Water supply and Water-mains........................... 114
2. Water-meter and By-pass................................ 116
3. Shut-offs and Gate-valves............................... 117
4. Fire-pumps... 118
      Details of Underwriter Fire-pump................... 121
      Fire-pump for Theatres............................. 124
5. Suction-reservoir...................................... 127
6. Fire Stand-pipes....................................... 128
7. Fire-valves.. 130
8. Fire-hose.. 132
9. Fire Nozzles and Couplings............................. 135

		PAGE
10.	Monitor Nozzles	136
11.	Hose-racks	137
12.	Hose-spanners	138
13.	Automatic-sprinkler System	139
14.	Perforated-pipe System	153
15.	Fire-pails	155
16.	Casks	159
17.	Portable Fire-extinguishing Apparatus	159
18.	Steam-jets	161
19.	Sand	161
20.	Fire-axes and Fire-hooks	161
21.	Minor Fire-extinguishing Devices	162
22.	Life-saving Appliances	162
23.	Fire-alarm and Watchman's Clock	163
24.	Water-supply for House Service	163

IV. LITERATURE ON THEATRES.

1.	Books—English	165
	" German	165
	" French	168
	" Italian	168
2.	Pamphlets	168
3.	Articles in Journals and Magazines	171
4.	Rules and Regulations	175
5.	Official Reports	175

THEATRE FIRES AND PANICS: THEIR CAUSES AND PREVENTION.

I.

THEATRE-FIRE CATASTROPHES AND THEIR PREVENTION.*

THE prevention of fire and incidental panic in halls of audience requires that in the planning, construction, interior arrangement and equipment of such buildings certain rules should be followed, and certain precautions observed by the architects and engineers of such structures. Of the greatest importance in respect

---

* This paper was prepared by the author for the annual meeting of the International Association of Fire Engineers, held in Montreal, Canada, in August, 1894. It was introduced with the following words: "*Gentlemen of the International Association of Fire Engineers:* Yours is an organization which has for its chief object 'the intelligent protection of human life and the saving of property.' Your annual meetings are held largely with the view of learning of improved methods of fire-fighting. Your principal business is to fight and put out fires, yet I believe I do not err in stating that *fire prevention* is a subject in which you are likewise interested. Indeed, the law in my own city, to quote an example, declares that 'the fire department shall be charged with the duty of *preventing* and of *extinguishing* fire and of *protecting* property.' You will observe that preventing fire comes first in the

to the enactment of strict building laws are all buildings where a multitude of people are congregated, by day or at night, for worship, education, or amusement, such as churches and synagogues, schools and colleges, concert- and lecture-halls, circuses and theatres. I shall restrict my remarks to theatres, although much of what I will say applies equally to the other classes of buildings named. The prevention of theatre-fire calamities is a subject in which not only firemen but the general public are, or ought to be, interested.

In the planning and construction of theatre buildings, even more than in other buildings, the combined skill, knowledge, and talent of both the architect and the engineer are necessary to create a model and safe structure. While the architect works out the best plan on the available site, determines the general arrangement of the divisions of the building, decides the mode of construction, designs the elevation, and selects the decorative treatment of the interior, he is assisted by the engineer or by specialists in the various engineering branches in the solution of intricate

---

enumeration of the duties of firemen. As in medicine there are two views, the one being the cure of disease, and the other one, which I consider the higher aim, the prevention of disease; just so, I hold, are fire departments and firemen concerned not only with the extinguishment of fires and with the saving of life and property, but likewise, and to even a higher degree, with the prevention of fires.

"About fire-fighting it would be presumptuous on my part to speak to you, for that is not in my line, and I would not be qualified to speak on the subject to you, many of whom have spent the best part of an active life in the battle with fire and smoke. The subject which I have chosen for my paper concerns the prevention of fire and calamities incidental thereto in a class of public buildings where large crowds gather nightly for amusement's sake."

problems requiring special expert knowledge, such as the construction of difficult foundations, the calculation and the working out of details of all iron constructions, the design of the mechanical or power plant, comprising steam-boilers, engines, dynamos, hoisting and general stage machinery, the heating and ventilation of the building, the gas-lighting and the electric-lighting system, the plumbing, drainage, and water-supply, the arrangement of the toilet-rooms, the general sanitation, and finally the fire protection of the theatre.

It is because I have had the good fortune of being associated in my own branch of engineering with architects of the highest standing in the construction and equipment of·some recent theatre buildings, and also because, by inclination, I have, for years, made the question of the safety of theatres a special study, that I venture to publish the following remarks about theatre-fire catastrophes and their prevention, in the hope that they will be helpful in preventing the recurrence of disasters similar to those related herein.

Theatre-Fire Statistics.

It may be well to begin by giving briefly a few statistics regarding theatre fires and their causes. Here we at once encounter much difficulty in gathering and presenting accurate and reliable figures and facts.

There are numerous cases in which a fire breaking out in a theatre, on the stage, in the auditorium, or elsewhere, is at once extinguished by the stage hands, or by the firemen on duty in the building. Many cases of this kind never become known to the public or to the press, and no accurate record is kept of them. In other instances, again, blind fire-alarms are followed by a panic and often by loss of life. When a fire in a theatre breaks out during the night and destroys the building the cause of the same often forever afterwards remains a mystery. Distinction must be made between occasional outbreaks of fire which are subdued before it can spread, and such fires which completely destroy the building. The data given below refer only to theatres which are entirely burnt down.

Those fires which break out during a performance, when the building is crowded with people, are naturally the ones of greatest interest to us. If accompanied, as they unfortunately often are, with loss of life, the number of persons killed and wounded is difficult to ascertain with absolute accuracy, as many perish in the building who are not missed by friends or relatives. The tendency in these cases is usually for the press to exaggerate the number of victims, whereas on the other hand the theatre officials or the municipality often try to keep the figures below the actual number. The curious mistake is sometimes made of counting both the dead and the wounded in the same figure of victims, obtaining thereby largely increased and incorrect figures.

We are principally indebted for theatre-fire statistics to the works of Herr August Foelsch, a German civil engineer, recently deceased; of Herr Franz Gilardone, a retired German fireman; of Capt. Shaw, of the London Fire Brigade; to Dr. Choquet, physician to a French life-insurance company, and to the labors and reports of John C. Hexamer, Esq., a civil engineer and insurance surveyor, of Philadelphia. Their figures and statistics do not by any means agree, and it is difficult to account for the discrepancies. Taking the work of Herr Foelsch as, on the whole, the most complete and reliable guide,—although it is not brought quite up to date, being published in 1878, with a supplement issued in 1882, shortly after the Vienna Ring Theatre fire,—we find a list of 516 theatres enumerated, which up to and including the year 1877 were completely destroyed by fire.*

Out of these, 460 theatres were burnt in 100 years. The list contains of theatres in the principal cities the following:

London......... 31	Boston........ 11	Venice......... 6
Paris........... 29	Glasgow....... 11	Baltimore...... 6
New York†..... 26	Cincinnati..... 9	Cologne....... 5
San Francisco... 21	New Orleans... 8	Edinburgh..... 4
Philadelphia.... 17	Bordeaux...... 7	

Among the 516 theatres we find—

37 theatres which were burnt twice.
8 "   "   "   "   " three times.
4 "   "   "   "   " four times.
1 theatre (the Bowery Theatre in New York) which was burnt ~~five~~ times; also one theatre in Spain, which burnt down seven times.

6 by 1929

---

* This figure in the supplemental edition to the work of 1882 was corrected to 523.
† Since 1821 up to the present time 29 theatres have burnt down in New York City.

Regarding the average life of theatres, the records are confined to 252 theatres, and show that—

5 theatres burnt down *before* the opening;				
70 " " " in the first 5 years after opening;				
38 " " " between 6–10 " " "				
45 " " " " 11–20 " " "				
27 " " " " 21–30 " " "				
12 " " " " 31–40 " " "				
20 " " " " 41–50 " " "				
17 " " " " 51–60 " " "				
7 " " " " 61–80 " " "				
8 " " " " 81–100 " " "				
3 " " " after over 100 years' existence.				

Total.... 252

Out of a total of 252 theatres, 70, or more than one fourth of them, are destroyed before they reach an age of 5 years. Theatres, therefore, as a rule, do not attain an old age, and Herr Foelsch figures the average age of these 252 theatres as $22\frac{3}{4}$ years. For the United States the average life of a theatre is said to be only from 11 to 13 years, but it should be remarked that these figures were reached some years ago, before the stricter theatre laws of some of our large cities (New York and Boston, and quite recently Brooklyn and Philadelphia) were enacted.

Recent analytical studies of the question have elicited the fact that there are two periods in the life of a theatre building, when it is most endangered or subject to destruction by fire, namely, one from the time of its construction up to and including the first five years, and the second period, after the theatre has been opened from 40 to 50 years. This may be explained, first, by the fact that in a new theatre the safety appliances, the arrangements for lighting, the scene-hoisting devices, the fire-protection appliances, are rarely in perfect working order, and the theatre employés have not as yet become accustomed to handling them, and are likewise unfamiliar with the rules of management.

The second period of renewed danger is explained by the fact that after the number of years quoted above much of the apparatus in a theatre has become worn out or useless, and is not always kept in repair or replaced by new apparatus, and it may also be attributed to the fact that after the lapse of many years the theatre management and inspection are apt to be less strict, and that often interior alterations are made which unfavorably affect the safety of the building.

From the middle of the last century to our present day nineteen theatres on the average have been destroyed annually by fire. For those who are interested in more detailed figures, I quote the statistics gathered by Herr Foelsch and Dr. Choquet, the latter's figures being placed in brackets.

In the years 1751–1760 there were 4 theatre fires in all countries.
" " 1761–1770 " 8 " " "
" " 1771–1780 " 9 (11) " " "
" " 1781–1790 " 11 (13) " " "
" " 1791–1800 " 13 (15) " " "
" " 1801–1810 " 17 " " "
" " 1811–1820 " 16 (18) " " "
" " 1821–1830 " 30 (32) " " "
" " 1831–1840 " 25 (30) " " "
" " 1841–1850 " 43 (54) " " "
" " 1851–1860 " 69 (76) " " "
" " 1861–1870 " 99 (103) " " "
" " 1871–1880 " 181 (169) " " "
" " 1881–1885 " (174) " " "

Taking each year from 1871 we have the following numbers:

In 1871	there were	20	theatres destroyed.	
" 1872	"	13	"	"
" 1873	"	15	"	"
" 1874	"	15	"	"
" 1875	"	14	"	"
" 1876	"	19	"	"
" 1877	"	17	"	"
" 1878	"	20	"	"
" 1879	"	25	"	"
" 1880	"	23	"	"
" 1881	"	28	"	"
" 1882	"	25 (37)	"	"
" 1883	"	22	"	"
" 1884	"	9 (40)	"	"
" 1885	"	9	"	"
" 1886	"	7 (8)	"	"
" 1887	"	12 (18)	"	"
" 1888	"	15	"	" (up to Nov.)

The fire loss represented by these figures reaches many millions of dollars' worth of property.

The hour of the day in which the above theatre fires started is of considerable interest. Out of 289 fires—

56	theatres	burnt	during the day between 7 a.m. and the afternoon, or.......	19.4	per cent.
15	"	"	one hour before beginning of performance, or..........	5.2	" "
36	"	"	during performance, or.......	12.4	" "
66	"	"	within two hours after close of performance, or............	23.9	" "
113	"	"	during the night and before 7 a.m. next morning, or......	39.1	" "

Later on, when the records of 373 theatre fires were available, Herr Foelsch found this percentage to remain nearly the same, viz.: 19.9, 5.6, 11.6, 22.6, and 40.3 per cent, respectively. It will be seen from these figures that the greatest danger from fire to a theatre is during the two hours following a performance, and not during the performance, as would naturally be supposed. The reason for this is partly, that during the performance greater watchfulness exists as regards open lights, the sources of heat, and the other usual causes of fire, and partly because many fires while actually started during a performance, for instance, by carelessness in the use of fireworks, or by the use of firearms, do not break out at once, but smoulder for a while in the inflammable scenery and woodwork of the stage, and break out during the hours following the performance. The risk from fire immediately before the performance and while the audience is admitted is found to be three times as great as during other hours of the day, which is explained by the fact that at this time the gas flames are lit which illuminate the scenery.

Theatres are, therefore, safest in the day-time; the danger is increased threefold during preparations before the performance, because of lighting up, etc.; it is reduced during the performance on account of the careful inspection and greater watchfulness on the stage, but is still two times as large as during the day; the danger reaches a maximum (7 times the day risk) during the two hours after the close of the performance, and it remains during the night nearly $3\frac{1}{2}$ times as great as during the day (sparks from fireworks or pistol wads may remain glowing for hours after the performance).

The theatre fires which break out during the performance are often accompanied with loss of life of both spectators and actors, and are therefore of particular though sad interest to us. The life of firemen is often greatly endangered at theatre fires at all hours, and the list of instances where firemen have been killed while engaged in their duty in trying to save burning theatres is quite large.

Herr Foelsch describes in detail 36 theatre fires between 1770 and 1877 which began during the performance, and out of this number not less than six were accompanied by appalling loss of life. From 1877 to 1889 six other deplorable theatre calamities have taken place. In the twelve years from 1876 to 1888 not less than 1600 people were killed in the six well-known theatre disasters of Brooklyn, Nice, Vienna, Paris, Exeter, and Oporto, nearly all of the victims being dead within ten minutes from the time when the flames and the smoke from the stage reached the auditorium and the galleries.

Dr. Choquet, in his statistics, enumerates a total of 732 theatres destroyed by fire from 1751 to 1885, with a total number of victims, i.e., both killed and wounded, of 6573. This list includes fires where firemen lost their lives in the performance of their duty. Since the beginning of this century 536 theatre fires occurred, and from 4500 to 5000 persons were killed or wounded.

Here are in detail, in periods of ten years, some of the figures of victims of theatre-fire calamities given by Dr. Choquet:

In the years 1751–1760,   10 victims.
"   "   1761–1770,   4   "
"   "   1771–1780,   154   "   (Amsterdam, 1772, 25 d.;
                                      Saragossa, 1778, 77 d., 52 w.)
"   "   1781–1790,   21   "
"   "   1791–1800, 1010   "   (Capo d'Istria, 1794, 1000 v.)
"   "   1801–1810,   37   "
"   "   1811–1820,   85   "   (Richmond, Va., 1811, 72 v.)
"   "   1821–1830,   105   "   (Philadelphia, 1829, 97 v.)
"   "   1831–1840,   813   "   (St. Petersburg, 1836, 800 v.)
"   "   1841–1850, 2114   "   (Canton, 1845, 1670 v.;
                                      Carlsruhe, 1847, 63 d., 203 w.;
                                      Quebec, 1846, 200 v.)
"   "   1851–1860,   241   "   (Livorno, 1857, 43 d., 134 w.)
"   "   1861–1870,   104   "
"   "   1871–1880, 1217   "   (Ahmednuggur, 1878, 40 v.;
                                      Brooklyn, 1876, 283 v.;
                                      Shanghai, 1871, 120 v.;
                                      Sacramento, 1876, 110 v.)
"   "   1881–1885,   628   "   (Nice, 1881, 70 v.;
                                      Richmond,(State?) 1885,100 v.;
                                      Vienna, 1881, 450 v.)
In the year 1887, Paris,   150   "
"   "   1888, Oporto, 140   "
"   "   1887, Exeter, 200   "
"   "   1892, Philadelphia, Central Theatre, several dead and
                    wounded.

Before proceeding to study the causes of theatre fires, I will give a brief list of twelve prominent theatre-fire calamities of this century, the horrors of many of which are probably fresh in the minds of some of my readers.

List of Twelve Prominent Theatre Fire Calamities of this Century.

1. Theatre in Richmond, Va.

Date: December 26, 1811.

Time of fire: During the last act of the evening performance.

Number of people in audience: About 600.

Cause of fire: Careless hoisting of a stage chandelier with lighted candles, whereby a border became ignited. Panic, jam at exits.

Number of victims: Seventy persons killed, many injured.

No data available as to location, plan, construction, and equipment.

2. Lehmann Theatre and Circus at St. Petersburg, Russia.

Date: February 14, 1836.

Time of fire: During the afternoon performance, at 4 o'clock.

Number of people in audience unknown.

Cause of fire: Stage-lamp, hung too high, ignited the stage roof. Panic, jam at exits.

Number of victims: About 800 persons killed.

Location: In an open square.

Construction: Temporary wooden structure.

Exits obstructed by the panic-stricken crowd.

3. Royal Theatre at Quebec, Canada.

Date: June 12, 1846.

Time of fire not stated.

Number of people in audience unknown.

Cause of fire: The upsetting of a lamp on the stage ignited a wing.

Number of persons killed: About 100.

Construction: Of combustible material.

Chief defects: Insufficient exits, narrow stairs; one wooden staircase broke from over-weight of the crowd, and thereby obstructed one of the exit doors.

4. Grand Ducal Theatre, at Carlsruhe, Baden, Germany.

Date: February 28, 1847.

Time of fire: Just before beginning of evening performance.

Number of people in audience: About 2000.

Cause of fire: The careless lighting up of the gas lights in the Grand Ducal box ignited the draperies.

Number of victims: 63 persons dead and 200 injured.

Location: Entirely free on an open square.

Construction: Wooden interior; the four exits ordinarily effected the emptying of the house in six minutes.

Chief defects: The main entrance to the theatre was bricked up (!). Of the four exits mentioned only ONE was open; the other three were locked and barricaded. The gas was turned off in the street after the fire broke out, but the oil-lamps in corridors were kept burning. The occupants of the parquet and of the balconies escaped. Most of the victims were gallery spectators, who became suffocated by smoke. Many were saved by breaking in the locked exit doors.

5. *Teatro degli Aquidotti, Leghorn, Italy.*

Date: June 7, 1857.

Time of fire: About 8 o'clock in the evening.

Number of people in the audience: About 3000.

Cause of fire: A rocket, part of fireworks on the stage, ignited the scenic decorations.

Number of victims: From 43 to 100 killed, and 134 to 200 injured.   No details.

6. *Conway's Theatre, Brooklyn, N. Y.*

Date: December 5, 1876.

Time of fire: During the last act of the performance of "The Two Orphans."

Number of people in audience: About 1000, viz., 250 in parquet, 350 in balcony, 405 in the gallery.

Cause of fire: A border caught fire from the border-lights, perhaps owing to a sudden draft, caused by opening a window. The fire was increased by the opening of a large door at the back of the stage.

Number of persons killed: 283, all from the upper gallery.

Location: Building stood detached on three sides.

Construction: Ordinary, but with well-arranged exits, permitting the emptying of the theatre in from 5 to 6 minutes.

Plan of theatre: Considered comparatively good at the time.

Chief defects: No fire-proof curtain, no water available for fire purposes. No fire-hose at the fire-hydrants, nor any other fire-extinguishing appliances available. Auxiliary exit doors for the gallery kept closed. Only one staircase for the gallery. Stage overcrowded with scenery. Loft over the auditorium filled with much inflammable scenic material. Proscenium of wood. During fire the gas was turned off in the street.

7. *Theatre Municipal, at Nice, Italy.*

Date: March 23, 1881.

Time of fire: At 8.30 in the evening, immediately before the beginning of the performance.

Number of people in audience not known.

Cause of fire: In lighting the border-lights, an explosion of gas took place, igniting at once the scenic decorations.

Number of persons killed: From 150 to 200, nearly all from the upper gallery.

Location: In an open square.

Construction: Substantial.

Plan and arrangement: Fairly good.

Chief defects: No fire-proof curtain, no oil-lamps in corridors and exits. Auxiliary exit door from gallery to the safety stairs could not be found on account of darkness, the gas-lights having been extinguished. The lower part of the gallery stairs steep, and with winding steps. The exit doors obstructed by the jam.

8. *Ring Theatre, at Vienna, Austria.*

Date: December 8, 1881.

Time of fire: About 6.45 in the evening, just before the beginning of the performance.

Number of people in the audience: About 1800.

Cause of fire: The careless lighting of the border-lights by means of an alcohol torch (the usual electric flash-light apparatus being out of order) ignited a hanging border.

Number of persons killed: At least 450, many from the upper gallery.

Location: Standing free on three sides.

Construction: Substantial, but stairways bad.

Plan and arrangement: defective.

Chief defects: Iron fire-proof curtain could only be lowered from the rigging loft, and could not be lowered at all at the time of the fire. Special exit doors locked, keys and locks rusty. Gas-lights extinguished, oil-lamps provided in corridors, but not lighted. The exits insufficient, and obstructed by the jam. This fire was very similar to the Brooklyn Theatre fire. The fire caught in each case among the flies, from a border-light. The bursting open of a large door in rear of stage let in a blast of air, which drove the flames and the suffocating fire smoke through the proscenium opening into the auditorium.

9. Circus Ferroni, at Berditscheff, in Russian Poland.

Date: January 13, 1883.

Time of fire: During evening performance.

Number of people in the audience: About 600.

Cause of fire: A stableman working in the stable adjoining the circus was smoking a cigarette and was lying on straw which ignited; a fellow laborer ran for a pail of water and left a door open, which created a strong draft and thus fed the flames. Inside of twenty minutes the whole circus stood in flames.

Number of victims: Official figure 268 persons

burnt to death; other reports state 430 dead, 80 mortally injured, 100 persons missing. Many children crushed and suffocated in the jam of people. Both stable laborers who caused the fire burnt to death; 27 horses and 11 trained dogs burnt.

Construction of building: Of double walls of wood, with an interior filling of straw (!) between the boards as a protection against the cold. No safety measures against fire. One safety exit, but not known to the public. Owing to the extreme cold weather there was scarcity of water, and a fire engine, in running to the fire, was delayed by breaking through the ice.

10. Opera Comique Theatre, Paris, France.

Date: May 25, 1887.
Time of fire: About 9 P.M.
Number of people in audience. About 1600.
Cause of fire: Scenery ignited from the gas-lights.
Number of persons killed: From 70 to 110, largely from the upper gallery, and stage people.
Location: Standing free on three sides.
Construction: Good, although not fire-resisting.
Plan and arrangement: faulty.
Chief defects: Exits for about half the stage people over a narrow wooden bridge over upper part of stage. Iron curtain not lowered; fire-hydrants not used; gas extinguished in the exits and corridors.

11. Exeter Theatre, Exeter, England.

Date: September 5, 1887.
Time of fire: During evening performance.
Number of people in audience not stated.

Cause of fire: Scenery caught fire from the gas-lights.

Number of persons killed: From 166 to 200, mostly from the upper gallery.

Location: In an open square.

Construction: Bad.

Plan and arrangement: Fairly good, but the upper gallery had only one exit, which was badly arranged and obstructed.

Chief defects: No fire curtain; insufficient fire appliances on stage; stage overloaded with scenery; insufficient and badly arranged exits.

12. Theatre at Oporto, Portugal.

Date: March 31, 1888.

Cause of fire: A rope in the rigging loft came too near the border-lights and caught fire, which quickly spread. A panic broke out, and the safety exits being closed a jam resulted in the corridors. Many persons jumped to the street from the windows, others followed on top of those lying wounded and killed in the street. Sailors and marine soldiers in the upper gallery are said to have used their knives to kill persons standing in their way.

Number of victims: 240 persons.

Chief defects: Theatre built entirely of wood; all stage gas-lights were open and unprotected lights. No safety appliances of any kind.

It may be noted that in nearly all the catastrophes briefly described fire broke out high up in the inflammable scenery near the borders, and was caused by gas-lights coming into contact with the combustible scenery.

Besides the twelve theatre fires briefly mentioned, in which alone about 3000 people lost their lives, there is a large list of other fires, full of horrors, about which, however, the details are meagre.

Among these I mention the fire in a theatre at Canton, China, on May 25, 1845, where 1670 persons are said to have perished; the fire of the theatre at Whampoa, China, in 1853, caused by fireworks, and killing 30 persons; the theatre at Tientsin, China, burnt in May, 1872, with 600 persons dead; the fire at a theatre in Ahmednuggur, East India, on May 11, 1878, when 40 persons were killed; the fire of the Gayety Theatre, in Milwaukee, on November 15, 1869, caused by the upsetting of a kerosene lamp on the stage, whereby 2 persons were burnt to death, and about 30 injured; the fire in a theatre at Tschernigow, Russia, on December 24, 1882, caused by a gas-meter explosion, and resulting in a panic, wherein several people were killed and 100 injured; the theatre fire at Dervio, Como, Italy, on June 25, 1883, where 47 persons lost their lives; and the fire at the Central Theatre, in Philadelphia, Pa., in 1892, causing the death of several persons.

Again, there are numerous instances of large loss of life during panics in public halls and places of assembly, of which I will mention only the panic at the Victoria Hall, in Sunderland, England, where, of 2000 children present, 183 were crushed and trampled to death; and the panic in a circus at Richmond, (State?) in 1885, whereby 100 persons were killed.

Causes of Theatre Fires.

Before discussing the precautions necessary for prevention, let us inquire briefly into the chief causes of theatre fires and calamities. This study of the causes leading to fires and panics is of the utmost importance, because it enables us to derive therefrom the best practical lessons as to what to do, and what to avoid, to prevent the occurrence of a disaster.

Dr. Choquet, in discussing the causes of theatre fires, divides them into two principal groups, viz.:

1. The intentional causes, or incendiarism; and
2. The accidental causes.

Leaving incendiarism, from malice or other reason, out of consideration, we can distinguish further between—

1. Exterior causes, such as lightning, exposure, dangerous trades carried on in the vicinity of a theatre, bombardment of the city, or riots.
2. Interior causes, viz.:
    a. On the stage;
    b. In the auditorium;
    c. In other parts of the house.
3. Mixed causes, such as fires originating in living apartments, or in stores, or offices, or restaurants, under the same roof with the theatre proper.

Placed in the order of their frequency, we find theatre fires to have their origin—

1. On the stage;
2. In the actor's dressing-rooms, in the engine and boiler-room, in the cellar, and in the administrative offices;

3. In stores, offices, living apartments, or restaurant kitchens in the theatre;

4. In the auditorium;

5. In outside causes, such as exposure to neighboring buildings on fire.

In not a few cases the exact cause of a theatre fire remains forever unknown. This is due largely to the rapidity with which the flames spread among the vast mass of combustible material, and to the haste and confusion incident to the fire, particularly when this breaks out during the night-time. The best information about causes of theatre fires is derived from those fires which break out during a performance, and which, for obvious reasons, are the most dangerous, and often lead to deplorable calamities.

The majority of fires during performances break out on the stage, and are due to open and unprotected, or to deficiently protected, lights in too close proximity to, or amidst a mass of, unprotected and highly inflammable scenery, draperies, gauze, ropes, and woodwork. Among numerous other causes I mention:

The careless lighting of gas-lights on the stage, by alcohol-lamps on long poles, and sometimes defects in the electric spark or flash-light apparatus, causing escapes of gas or gas explosions;

Carelessness in the use of lamps on the stage or in the dressing-rooms, or exposure of lamps to sudden drafts;

Careless handling of candle-lights in dressing-rooms;

Alcohol-lamps used by actresses for heating hair-curling irons in the dressing-rooms;

Sparks dropping from torches used to light the gas flames on the stage, or on decorations or papers;

Temporary gas conduits needed on the stage, generally consisting of rubber tubing, becoming leaky, or not being well jointed together, or becoming otherwise damaged and defective;

The use of portable open gas-burners near movable scenery;

The use of defective gas-burners and gas-fittings;

Exposed gas-pipes damaged by the shifting or carrying about of heavy pieces of scenery or furniture;

Main gas-pipes becoming leaky, and causing escapes of gas;

Gas explosions in the gas-meter vault, or elsewhere;

Swinging gas brackets setting woodwork or scenery or curtains on fire;

Lights other than gas-lights are often worse; the upsetting of coal-oil or kerosene lamps and the explosion of such lamps are fruitful causes of fire.

The electric light, finally, although the safest comparatively of all lights, may cause fire by reason of electric-light wires not being properly insulated, or being faultily installed.

The careless use of matches; other than safety-matches, throwing away of lighted matches, and rats or mice gnawing phosphorous matches in hidden recesses, are among the causes of fires;

The smoking of tobacco, cigars, cigarettes, as well as pipes, and the careless throwing away of lighted cigars or cigarettes in dressing-rooms or on the stage or in the sub-basement;

The careless use of the oxy hydrogen or lime light;

The use of fireworks in spectacular plays calculated to attract the public; the use of red fires, torches, Roman candles, the firing of rockets,—are dangerous causes of fire, because a spark often lodges amidst scenery, and the flames break out during the night following the performance;

Sometimes fireworks are prepared for the play in the theatre building, which is a pernicious practice;

The representation of fire or conflagrations or actual explosions in the course of the play, the firing of fire-arms, such as pistols, or guns, with paper wads;

The careless handling and use of benzine or turpentine in workshops and costume-rooms;

The heating-apparatus, particularly stoves and furnaces when overheated during extreme cold weather;

Defective flues, and timber near woodwork;

Floor registers and heat flues; hot ashes;

The use of straw or hay on the stage;

The careless use of soldering furnaces on tin theatre roofs;

Gauze dresses of ballet-dancers catching fire and in turn setting decorations on fire;

Non-removal of rubbish and trade refuse, and accumulation of large masses of combustible material, woodwork, canvas, gauze, pasteboard, hemp ropes, paper with varnish, so-called properties, etc., on the stage;

The use of rooms in the theatre, in the **basement or** in the loft over the auditorium, or even of **the stage**, as painter's or carpenter's shop;

The spontaneous combustion of heaps of waste

material, of oily rags used for cleaning lamps, or of waste in engine or dynamo room;

The action of rays of the sun concentrated by window glass or lenses;

Carelessness in the necessarily quick moving or shifting of inflammable scenery near the innumerable gas-lights;

The drying up of woodwork on the stage by the heat due to the numerous lights.

As *indirect* causes of fire I may, lastly, mention:

An undesirable site, the danger from contiguous or adjacent houses and property;

The bad arrangement of the plan;

The use of bad materials in construction;

Defective general system of construction without division into separate fire risks;

The misuse of a part of the premises, as living or sleeping apartments, as kitchens for restaurants or for dangerous trades, or stores;

The want of proper fire appliances and of a fire watch;

The absence of fire-alarm telegraphs, or delays in sending the alarm;

Frequent changes of watchmen, fire watch, and stage hands, and the fact that even men occupying responsible positions grow careless after long service.

The danger of fire in a theatre is also, to some extent, dependent upon the character of the play, whether the plain comedy or drama, or the more elaborate opera, ballet, or spectacular play; upon the arrangement and size of the stage, the method of lighting used, incandescent electric lights being without doubt the safest to use; and the age of the building, the first five years being, as found by experience, among the most dangerous.

Causes of Panics.

Panics may occur in any place of assembly where large crowds of people are congregated, even when there is no outbreak of fire, and sometimes very trifling causes may create terrible panics. It is in many respects much more difficult to prevent a panic in a theatre than a fire; and, what is worse, panics may become quite as dangerous in fireproof buildings as in those of a combustible nature.

The immediate causes of a panic may be:—

A sudden outbreak of fire;

A false cry of "fire," the false alarm being given either by some excited or frightened, foolish person in the audience, or sometimes intentionally, as for instance by pickpockets;

A sudden commotion on the stage or in the audience;

The sudden appearance of smoke on the stage, from burnt fireworks or otherwise;

The unannounced darkening of the house;

A sudden extra pressure of gas at the footlights, causing the flames to flare up unduly;

The sudden extinguishment of the electric light;

A fire-alarm in the neighborhood, or even only the rattling noise of passing fire-engines;

The reflection of a fire in the neighborhood of the theatre;

A thunderstorm or stroke of lightning;

Persons fainting in the theatre;

Horses on the stage becoming frightened and shying, or falling into the orchestra pit;

And miscellaneous other causes.

Features in the construction and arrangement of a theatre which may indirectly lead to a panic are:

Too few exits;

Various exit passages crossing each other;

Obstructed aisles or corridors;

Too narrow stairs, or winding stairs;

Doors opening inward, and doors opening in dark corridors;

Safety exits kept under lock and key.

The best way to prevent panics is to give to the theatre-goers a feeling of security by planning a very ample number of wide exits, so as to enable the quick and quiet emptying of the house; by arranging easy stairs; by providing at least two separate stairs for each of the tiers; by arranging a larger number of stairs for the visitors in the gallery; by installing the electric light in the theatre, particularly on the stage; and providing also auxiliary lights in the corridors, stairs, and exits. Furthermore, by making it known to the theatre-goers that safety appliances are provided in abundance, chief among these being large stage ventilators, a fire-proof curtain, fire-walls dividing the stage and the auditorium, the fire-proof treatment of scenery and of woodwork on the stage, also the putting up and keeping in readiness, for instant use in an emergency, of plenty of fire-extinguishing and life-saving appliances, including an automatic sprinkler system on the stage, which must be kept in good working order. Finally, by informing the public that a strict fire watch is kept, that the theatre is frequently inspected, and by issuing and publishing stringent regulations referring to theatre management.

Dangers to Human Life.

To further understand correctly the subject of theatre-fire catastrophes, let us consider briefly the various ways in which human life is endangered in a theatre fire or panic.

At the outset I will state that it is an error to suppose that the flames are the chief cause of death in theatre-fire catastrophes. The lessons of all theatre calamities point to the fact that it is not the fire, but the smoke, the fire gases, and the intense heat incident to the fire, and on the other hand the mad rush in a panic, which kill most people. Indeed, the majority of people, particularly in the upper galleries, are suffocated or trampled to death before the fire reaches them. In many of the theatre conflagrations cited above, the bodies of victims were found perfectly intact, and either in a sitting position or else fallen upon the floor dead, without the flames ever having reached them.

The lives of people in theatres, whether spectators, actors, musicians, chorus-singers, ballet-girls, or stage hands, are therefore endangered—

First, by smoke, fire gases, heat, asphyxia, exhaustion;

Second, by the flames of the fire;

Third, by jams, knocking over, falling down stairs, trampling, crush;

Fourth, by direct shock or fright;

Fifth, by accidents, such as the falling of the central chandelier.

These various causes may result in:

(*a*) Almost instantaneous death;

(*b*) Mortal injuries;

(*c*) Injuries resulting in permanent bodily disablement;

(*a*) Temporary illness and recovery.

The long list of theatres destroyed by fires breaking out during a performance, and the numerous instances of fires breaking out during these hours, but which are put out before spreading, is proof sufficient that the dangers spoken of are constantly threatening the theatre-going public.

To sum up: Chief among the causes to which serious calamities are due are: open lights and inflammable scenery; defects in gas-lighting and gas-piping; defective gas-lamps; faulty construction of the building; deficient exits and absence of a fire-proof curtain, or, where same is provided, the fact that the curtain is not used, or is out of order; absence of fire-extinguishing appliances, or lack of water under fire-pressure; auxiliary lights in corridors, exits, and stairs not; provided absence of smoke ventilator over the stage; accumulation of inflammable scenery, and bad management in the theatre.

We have now gained a sufficiently clear understanding of the causes of fires and panics in theatres to enable us to turn to the second part of our subject, viz.:

The Prevention of Theatre Fire Catastrophes.

The principal measures for the prevention of theatre-fire disasters should have in view, first, the safety of the people in the theatre, whether spectators, actors, stage hands, or firemen on duty, and next, the safety of the building. The safety of the people comes first in the order of importance. The building may even burn down, but the public must under all circumstances have facilities for saving themselves. All efforts must be directed towards making the play-goers, the players, and the theatre employés safe. All manner of danger arising from a fire or from a panic, caused by a real or false fire-alarm, should be averted, and appliances for saving life should be kept within reach. Next, the building should be made structurally safe, and efficient means for fire extinguishment should be provided, thus preventing the immense fire loss usually caused by a theatre conflagration.

The safety measures to be briefly considered hereafter may be divided into four principal groups, viz.:

(*a*) *Measures to Prevent Outbreaks of Fire* which include rules of management, regular systems of inspection; considerations of plan, size, location, and construction of a theatre, height of building and number of galleries, and structural details.

(*b*) *Measures to Localize Fires* and to prevent their spreading, if in spite of precautions they do break out. Under this heading would come the fire-walls, the fire-doors, the fire-curtain, the floor, and roof construction, and the division of the theatre, horizontally and vertically, into many separate fire risks.

(c) *Measures to Insure the Safety of the Spectators and of the Stage People*, to prevent injuries resulting from a jam or crush and a panic, the suffocation by smoke and death by fire. This group includes ample means for safe, easy, and quick exit from the burning theatre, and relates to such subjects as seats, aisles, passages, doors, stairs, and exits. It also embraces the questions of lighting and of ventilation, provision of stage ventilators, and special safety appliances, such as thermostats, automatic fire-alarms, etc., telegraphic communication with the fire-department, etc.

(d) *Measures to Put Out Fires*, which include all fire appliances; the stage-sprinkler system, etc.; a good, plentiful water service, etc.

The whole subject will be discussed more systematically and at greater length in the next chapter, on " The Essential Conditions of Safety in Theatres." It will suffice to enumerate here only the principal measures of safety.

1. To begin with, the theatre building should be isolated as far as practicable. Then there should be in every theatre plenty of wide, well-arranged and well-distributed exits, leading the audience as quickly as possible to outdoors, and thereby decentralizing the crowd. There should be separate and numerous fireproof stairs—at least two for each tier. For the stage and its accessory departments provision should be made for at least two independent exits. Fireproof stairs for the workmen should lead from the rigging loft

down to the stage floor. The fire at the Opera-Comique taught a good lesson in pointing out the necessity of having proper means of escape for the performers and workmen on the stage, which sometimes number as many as 300 to 400 people. Where the theatre is not standing free on all sides there should be wide courts on both sides and additional safety exits from the auditorium to the same, including suitable fire-escapes from the balcony and upper galleries. Much the best plan, where the dimensions of the lot permit, is to have wide fire-proof terraces, foyers, or colonnades adjoining the corridors in each tier, which form places of safety for the audience in case they have to leave the theatre suddenly. All halls, corridors, passages, stairs, exit doors, courts, and exits should be free from all obstruction. All exit doors should open outward. Exits should not be reduced at any point, but if possible should widen outward. There should be plenty of wide aisles, without steps, and no chairs or camp-stools should be permitted in the aisles.

It is particularly important that there should be a sufficient number of stairs and plenty of exits for the occupants of the galleries, who, as the fire catastrophes described amply prove, are more endangered than the other spectators. Those who pay twenty-five cents admission fee have surely the same rights to safety as the occupants of three or five dollar seats; and in a place of amusement, to quote the words of the Rev. Dr. Duryea, spoken at the funeral services for the victims of the Brooklyn Theatre fire, " the poor should be made as safe as the rich."

Stairs from different tiers should not communicate together, nor should streams of people cross each other on their way out. All stairs should be provided with handrails on both sides; winding stairs and single steps should be avoided.

The provision of proper exits (including, in the broader sense of the word, aisles, corridors, stairs, doors) is by far the most important safety measure both for fire and for panic. In fact, if properly carried out, it will save the lives of people even where all the other precautions are neglected. The aim should be to arrange the exits so that a theatre can be emptied without the slightest difficulty in from two to three minutes. Much will depend upon the available site and upon the plan of the building.

If only plenty of exits are provided, so that, *under all circumstances*, the whole audience, even when frightened and suddenly thrown into a state of high mental excitement, can leave the building inside of two or three minutes, the fire-resisting qualities of the building are of less consequence, as regards the safety of the persons in the theatre. In fact, a theatre inferior in point of construction, but having exits as above described, would be safer than one built thoroughly fire-proof, but otherwise not well arranged and not provided with sufficient stairs and exits, and where, therefore, in case of a false or real alarm of fire, or a panic from any cause, the people would necessarily be in grave peril.

2. Recognizing from the lessons taught by past theatre-fire catastrophes that the stage is the chief point of danger, it should be the aim to completely

isolate this part of a theatre. Fire-walls should separate the stage from the auditorium on the one hand, and from the dressing-rooms and theatre offices on the other. The fire-wall, dividing the auditorium from the stage, in which is located the large proscenium opening, should have as few other openings as possible, and none at all above the level of the stage. All openings should be kept closed by fire doors. The proscenium opening should be fitted with a fire-resisting curtain, closing as nearly as possible hermetically. The chief object of the curtain is to localize a stage fire and to cut off the fire, the smoke, and the view of the fire during the retreat of the audience. It thus serves to restore confidence, and will often help to avoid the usual panic and jam. The fire-curtain should be able to withstand the increased air-pressure due to the heat sufficiently long to enable the audience to make its escape. Whether the curtain be of corrugated iron or of asbestos, it is important that the apparatus for lowering the same be located on the stage, and not in the rigging loft, which place very soon becomes inaccessible when a stage fire breaks out.

3. Smoke and fire gases being the chief causes of death of the victims of theatre-fire calamities, it is essential that large smoke-ventilators or ventilating skylights be fitted up over the stage, in order to remove the smoke at the highest point of the stage roof. Incidentally, the stage ventilators tend to create a current of air from the auditorium towards the stage, and thereby prevent smoke, flames, and suffocating gases from escaping into the auditorium. In this connection the abolishment of the usual central chan-

delier and ceiling vent in the auditorium is important. All ventilating registers for the auditorium should have dampers which are controlled and can be closed from the stage. This measure will tend to prevent the suffocation of the people in the upper gallery in case of a fire.

4. The study of the causes of theatre fires having shown that open gas-lights in the vicinity of stage scenery and settings are the most prolific cause of fire, it is obvious that the stage and the scenery can be made safer by fitting the same up with incandescent electric light. All open lights, and gas-lights in particular, as well as all sources of heat, should be eliminated from the stage. Incandescent electric light constitutes a brilliant, uniform, and easily regulated light. It has not only the advantage of being much safer as regards danger from fire, but it is superior also because it does not vitiate the air, because it creates much less heat than gas, and because it does away with the dangerous lighting up of gas flames, which has been the cause of many a fire. In short, the electric light in theatres has all the advantages of gas without any of its inconveniences and dangers. But if gas-lighting must be used, the gas-piping should be most carefully done, all open flames should be well protected, the gas-meters set in a safe place, and the best available electric spark or flash-lighting system should be installed for lighting up. In addition to the regular system of lighting, whether by gas or by electricity, there should be provided in the corridors, stairs, and exits auxiliary lights, either candle-lanterns, or oil-lamps burning vegetable-oil, to make the retreat of the people safer

in case the gas is turned off or in case the electric light fails. The central gas chandelier in the auditorium, moreover, should be abolished.

5. To increase the safety of the stage still further, it is imperative to insist upon the fireproof treatment of all stage woodwork by the application of fire-proof paint, and likewise upon the chemical impregnation of all gauze scenery, draperies, furniture, and costumes to render same uninflammable. A further step forward, in the right direction, consists in the employment of light sheet-iron frames or of wire-cloth netting, or in the use of asbestos for scenic decorations. Improvements in the stage machinery, such as the use of hydraulic power for raising and lowering traps and scenery in place of wooden hoisting drums and hand labor, the substitution of steel-wire ropes in place of hemp cord, the entire abolishment of borders and wings, are much to be desired, all of which tend to make the stage safer. These are not merely theoretical suggestions: they have, on the contrary, proven eminently practical in actual use. The improvements recommended are largely due to suggestions from the Asphaleia Society of Vienna, Austria, whose system of stage construction and stage machinery has been successfully used at the theatres of Buda-Pesth in Austria-Hungary, at the People's Theatre in Vienna, at the Halle Stadt Theatre in Germany, in the Victoria Theatre in Sidney, Australia, and in our country in the new Auditorium Theatre in Chicago.

The accumulation of much scenery or scenic decorations on or near the stage should be avoided, and not more scenery should be hung than is needed for two performances. Neither the auditorium loft nor the cellar under the auditorium should be used for the storage of scenery. Want of order on the stage is sometimes due to lack of store-room, or to an insufficient number of stage hands.

6. In the construction of the theatre building, fire-resisting materials should be preferred, and unprotected ironwork, granite, and limestone should be avoided. The building should be divided into a large number of separate fire risks, both horizontally and vertically. The stage should be completely isolated by fire-walls carried above the roof, all dressing-rooms should be made fire-proof, and the whole auditorium and each of its tiers, corridors, and stairs should be constructed in a fire-resisting manner.

7. All scene-docks, workshops, carpenter and paint shops should be banished from the theatre building, and placed in a separate fire-proof annex. The heating apparatus, the dynamos of the electric-light plant, the fire-pump, and the gas-meters should be located in fire-proof vaults, preferably under the sidewalk, but not under the stage nor under the auditorium. No living or sleeping apartments, and no stores should be located in the theatre building, and no dangerous trades should be carried on in the same.

8. In order to put out a fire in a theatre before it can gain headway, there should be provided an abundant supply of water under fire-pressure, and plenty of efficient fire-extinguishing appliances, such as one or more fire-pumps, fire stand-pipes, fire-valves, hydrants, monitor nozzles, and fire-hose; a large number of fire-pails, casks of water, chemical fire-extinguishers, fire-

axes, fire-hooks on poles, wet sheets, wet sponges on long poles, etc. The whole stage should, moreover, be protected by an automatic sprinkler system, supplied from a large open tank on the highest roof, and having also an outside fire-department connection, to which the fire-engines may be coupled. All the appliances mentioned are useful chiefly during the first moments of a fire, and wherever provided and kept in good order and in readiness for instant use, they have served to put out many a stage fire which, without them, might have resulted in a serious calamity. It is true, however, that, as soon as the flames have spread, in the majority of theatres as constructed until recently, the fire appliances would soon become powerless to subdue a fire. In regard to details, I refer to Chapter III, which deals in particular with "The Water Service and Fire Protection of Theatres."

9. Much care should be bestowed upon the heating apparatus for a theatre, for in not a few instances has it been the cause of a fire. A number of fires as sources of heat being out of the question, on account of difficulties in attending to them, and also because they increase the fire danger, the choice lies between a warm-air furnace, a steam or a hot-water boiler. Furnaces can only be used for theatres of small size, and in the majority of cases heating is done by steam. The usual precautions should be observed, and the steam-boiler placed in a fire-proof vault, preferably outside of the theatre proper. It may with advantage be combined with the electric-light plant.

10. Efficient lighting-rod protection adds to the

safety of a theatre building, and in European theatres its installation is generally included in the equipment of the building.

11. It is quite important to have on hand, in every theatre, for cases of emergency, a few life-saving appliances, such as rope escapes, a jumping-net, and a cloth chute to aid in saving persons in case their retreat from the upper floors should be too suddenly cut off by fire or smoke.

12. Finally, it will be a great help in preventing theatre fires if a fire watch is kept in the building day and night, reinforced during the hours of the performance by detachments from the city fire-department. Nothing will increase more the security of a theatre than frequent inspections, about which I shall say something hereafter. There should be a telegraph connection with the nearest fire-department station, and automatic fire-alarms at many points in the building, which should be frequently tested to make sure that they are in good working order. There should also be speaking-tubes, electric bells, and telegraph alarms, to bring all parts of the building into communication with each other. And, lastly, the theatre employés should undergo regular fire-drills, so that in case of an emergency each employé will know what duty he has to perform.

Many of the safety appliances spoken of can be arranged so as to work *automatically*. Such automatic appliances are good in their way, and some of them have proven in actual experience to be quite efficient. They should not, however, be solely relied upon, and in general it will be better not to trust to automatic appliances altogether. The automatic fire-alarm system and the automatic sprinkler system may be approved; but the automatic sliding skylights or the smoke ventilators over the stage roof and the automatic fire-curtain may fail to work properly just when needed. I am quite aware of the fact that others argue in favor of automatic appliances, claiming that one cannot rely in moments of danger upon the cool-headedness of men, and that in a panic every one thinks of his own safety first, and that theatre employés will forget their duties.

But, in my judgment, it is infinitely better to put all such appliances in charge of a special, trusted, "safety officer," who may be an experienced fireman, whose duty in case of an outbreak of fire should be to lower the fire-curtain, to open the stage ventilators, to close the auditorium ventilating registers, to send the alarm to the theatre engineer who runs the fire-pump and to the nearest fire-engine station, to notify the audience promptly that they must disperse quietly, to see that the gas is not shut off and that the auxiliary lights are kept burning, to see that all doors leading to the stage are kept closed to prevent a draft, and who should order the water turned on at the fire-hydrants, the monitor nozzles, and the perforated pipe system, where such is installed in place of automatic sprinklers.

I have not, so far, made any distinction between *old* and *new* theatre buildings, for the requirements for safety really apply to both alike, but, of course, are much more difficult to enforce in the case of the older structures. These are, everywhere, as a rule, lamentably deficient and unsafe, and much good would result if they were frequently inspected by the fire-department, and if alterations were ordered made to increase the safety of audiences. The public generally is not able to, and does not, discriminate between safe and dangerous theatres. If the older theatres cannot be made safe, particularly as regards the exits, they should be closed up by the authorities. All theatre regulations should be compulsory, and the building, fire, and police departments should have power to stringently enforce them. The law should clearly define the responsibility of architects and builders and of the theatre managers in the matter of theatre safety. It is but reasonable to require of theatre managers that every known approved measure, tending to increase the safety of the public, be provided for. It is, likewise, justifiable to require builders to construct places for public amusement in such a secure manner that all dangers to life and limb are averted.

Theatre Inspections.

In the case of new theatre buildings, it does not suffice to have them well planned and well constructed. There should be, after the opening, regular inspections to make sure that the laws are not violated after the new building has passed the final examination of the authorities. If subsequent alterations are contemplated, they should likewise be made to conform to the building laws. The theatre license should be subject to revocation at any time for violation of the law. In the efficient periodical inspection and control of theatres by the authorities lies the greatest safeguard against fire catastrophes. Such inspections should be made much oftener than once a year. In Vienna they occur four times a year; in Paris inspections are made every month by a committee of safety, consisting of a police commissioner, an official from the city fire-department, and an architect. In London monthly inspections are required. These inspections should be made not only in day-time, but likewise in the evenings during a performance, and all details should be included in the examination. Special expert surveys and tests of the gas-pipes, electric conduits, fire-alarms, and the fire appliances should be made from time to time, and official reports made as to the results found. It is best to make inspections without any previous announcement. The results should be published, without fear or favor, in the daily newspapers. Nowhere are theatres inspections carried out with greater strictness than in Berlin. All theatres are regularly inspected and surveyed, at odd times, by a committee from the fire brigade and the building department. Every two weeks the district fire-marshal examines the theatres, during the day and also at night. Finally, officers of the fire brigade make nightly inspections during the performances, and a watch of trained firemen is placed in every theatre during performances. All these precautions have a tendency to awaken public confidence, and in case of a fire a panic is not so apt to occur. Indeed, there are several instances of well-built and well-managed theatres on record where during a performance fire broke out which ultimately destroyed the building, but where the whole audience left the theatre quietly and in good order, and where no accident of any kind occurred.

For the safety of theatres it is essential that they be continuously watched. In the words of Mr. Garnier, the architect of the Paris Opera-House, "the strict, minute, and incessant watch and inspection of all parts of a theatre constitute the chief defence of theatres against fires."

A century ago it was decided in France that firemen were the proper persons to do this. At first they were present on the stage merely during the performances; subsequently it was decreed that firemen should be on watch in a theatre during the day and the night. If the employment of fire-watches is left to the discretion of theatre managers, persons are sometimes engaged for this duty who are incompetent, or, if competent, they are required to perform other duties besides, and being thus overworked, fail to efficiently accomplish the object sought for. Fire-watchmen should be well acquainted with the building, the whole theatre staff should be under their control, and they should be vested with authority to interfere in case of violation of the theatre regulations. In the large Paris Opera-House there are always twenty-five firemen on duty, and during performances their number is doubled. In the Vienna Opera-House there are ten men on duty. In the Berlin theatres strong fire-watches, composed of the most experienced men of the fire-brigade, are stationed in the building during performances, and a special police patrol is stationed in front of the house to keep the crowd in order, and to see that the exits are kept open and unobstructed. During all performances a detachment of firemen should be stationed on the stage and should watch not only the lighting arrangements, the fireworks, the firing of fire-arms, but also have charge of the fire-extinguishing and life-saving appliances, and see that they are kept in order and ready for use. At the close of each performance an inspection of the whole theatre should be made by the fire-watch, attention being paid in particular to the heating and lighting apparatus, to the decorations and scenery, and to the dressing-rooms.

One cannot learn the dangers incident to large fires, nor the effect of good fire-extinguishing appliances, better than by being daily engaged in fighting flames, nor can anybody have a better appreciation of the effect of suffocating smoke than a fireman. Firemen, therefore, are as well qualified as builders, architects, and engineers to judge of the security of a theatre building and of the efficiency of the safety measures adopted. The training of firemen is such as to make them proper persons to examine and pass upon the plans for new theatres, and to inspect both—the theatres in course of erection as well as those already built. With their assistance the rules and regulations for theatre management should be framed; they should be intrusted with the inspection of theatres, to see that the rules are enforced; finally, they should be in charge of the fire appliances in theatres, and should direct the fire-drill of the employés.

Theatre Management.

Strict rules and regulations pertaining to the management of theatres should be drawn up by the authorities, and enforced by the fire, building, and police departments, which, as regards theatres, are correlated. Such rules of management relate to the following, viz.:

The proper storage of decorations, scenery, furniture, and property;

The maintenance of general order, cleanliness, and discipline;

The removal of rubbish, litter, and ashes;

Keeping the scenic decorations free from dust;

Keeping all corridors and exits unobstructed, and likewise holding all fire stand-pipes, hydrants, fire-pails, and casks of water accessible and ready for use.

The rules also relate to the use of open fires and lights, the employment of special lights and fire effects, the use of fireworks, and of fire-arms, the representation of actual fire scenes, the use of matches, the prohibition of smoking or lighting cigars in the auditorium, and foyers, or in the dressing-rooms, and should only permit the same where needed in the course of the play. They should also include the daily use of the fire-curtain, and the opening of all exits; the maintenance of auxiliary lights, either oil-lamps or candle-lanterns, in corridors, courts, and exits; the keeping up of steam under proper pressure at the fire-pump during performances, the employment of a special fire-watch and a theatre night-watch, etc. They should require the fire-proof treatment of all gauze costumes, and prohibit the use of open lights in the wardrobes and dressing-rooms. A penalty should be enforced for the use of fire-pails for other than fire purposes. Moreover, the number of persons in each tier should be limited by law, and standing-room in the aisles should be prohibited, and a heavy fine enforced for any violation of these rules. The practice of a sudden, unannounced darkening of the auditorium should not be permitted, nor any dangerous fire exhibitions on the stage, except where the stage is absolutely fire-proof. Special precautions should be enforced where the stage is decorated with natural branches of fir-trees, which when exposed to heat dry out quickly and become extremely fire-hazardous, and also where the auditorium and the stage are temporarily thrown together, as during masquerade balls. The rules should prohibit workshops of any kind in the theatre proper. They should finally require clearly drawn plans of the theatre, showing the exits, to be hung up at conspicuous points, and also to be legibly printed on all theatre programmes; and, last, they should compel the plain marking of all exits in large letters painted over the doors.

Conclusion.

Only a few more words in conclusion. Experience teaches that after each theatre fire, accompanied by loss of life, the greatest excitement prevails for a time, and the daily papers, the magazines, and the technical press are full of articles, discussing defects of theatres and suggesting remedies; the public refrains temporarily from going to these places of amusement;

the authorities show the greatest zeal in making official inspections; and theatre managers actively undertake interior improvements. Very soon, however, the excitement subsides, after a little while the lessons of the calamity are forgotten, the theatre management resumes old habits, the most common precautions are neglected, the public becomes again indifferent to the dangers constantly threatening them in unsafe theatres, and everything goes along as before. This, I need hardly say, is all wrong.

There is an old adage which says, " In time of peace prepare for war." Applied to our subject, it means that the movement for theatre reforms should constantly be kept stirred up, that, without unnecessarily alarming the public, interest in the subject of theatre-fire prevention should be ever maintained, and that, quietly but persistently, radical measures should be adopted to stamp out the grave dangers to which human life is exposed in the theatres of many of our cities.

I should feel amply rewarded for the labor involved in gathering the facts contained herein if it should result in the improvement of some theatres, and lead to the enactment of stricter laws regarding theatre construction and theatre management in many of our smaller cities and towns, where, even at this day, no official control over theatres is exerted by building or fire departments. Strict theatre laws and frequent inspections by fire-departments constitute the most important factors in lessening, for the future, the number of theatre-fire catastrophes.

II.

THE ESSENTIAL CONDITIONS OF SAFETY IN THEATRES.*

THE subject of the causes and prevention of fires in theatres has, for many years, occupied my special attention. Having been engaged in the course of my professional practice with the installation of water, gas, and fire service plants in some of the modern theatres, I have had special opportunities of becoming familiar with the essential requirements of safety in these buildings.

In the following an attempt has been made to give a brief *résumé* of these requirements, as derived from personal experience, observation, and study. As regards my views on the fire protection of theatres, the reader is referred to the detailed description given in Chapter III, which deals with "The Water-Service and the Fire-Protection of Theatres." The gas-lighting system, the plumbing, ventilation, and sanitation of theatres are likewise special subjects worthy of careful consideration.

The safety of a theatre may be said to depend on six principal factors, viz.: (1) The site; (2) the plan; (3) the construction; (4) the interior equipment and

---

* This essay on Modern Theatre Planning, Construction, Equipment, and Management was originally prepared for the *American Architect*, and appeared in the June and July, 1894, issues.

arrangement; (5) the management, and (6) the periodical inspection.

The *site* of a theatre building is of importance as regards its relation to adjoining buildings, and as regards the approaches, entrances, and exits. The development of the plan depends largely upon the available site.

The *planning* of a theatre consists of the arrangement of the floor-plans of the building, and of the various parts of the same, with their principal horizontal and vertical subdivisions and means of communication, such as stairways, entrances, and exits.

The *construction* of the building is to be considered in reference to the employment of fire-resisting materials in general, and also, in detail, in regard to the construction of the stage and its accessories; the auditorium and tiers; the dressing-rooms; workshops; the roofs; the aisles, staircases, passages, corridors, doors, entrances, and exits.

The *interior arrangement and equipment* include many important details, such as the fire-proof curtain, the stage scenery and stage machinery, the fly-galleries, rigging-loft and painter's-bridge; the stage-roof ventilator; the heating, ventilation, lighting, plumbing, drainage and sanitation, water-supply and fire-extinguishing appliances; lightning-rods; fire-escapes, doors, auditorium-chairs, and the life-saving appliances.

The *management* includes numerous matters of detail and rules and regulations pertaining to the safety of the building on the one hand, and of the audience, the actors, and the stage employés on the other.

The *periodical inspection*, finally, forms a necessary and important safeguard, and includes frequent surveys of the building and of its gas, water, steam, fire, and electric appliances, all of which should be tested at regular intervals, to insure their being in proper working order when needed.

Causes of Theatre Fires.

A study of the chief causes of theatre fires is important and necessary to a proper understanding of the subject. The knowledge of the principal causes to which theatre catastrophes have been attributed will do much to enable us to avoid, or to remove, defects which, if present or allowed to remain, may at any time precipitate a calamity. I must, however, content myself with giving the following brief summary:

(*a*) Bad location; exposure to fire from neighboring buildings.
Bad planning; faulty interior arrangement.
Inferior or improper construction; structural defects, defective flues, timber near flues.

(*b*) Wrong use of the premises; dangerous trades carried on in stores or shops in the theatre building.
Defective lighting apparatus; oil-lamps, gas and electric light; matches, torches, spirit-lamps; leaky gas-pipes, gas explosions.
Defective heating apparatus.
Accumulation of highly inflammable stage material.
Use of fireworks, colored lights, explosives,

and of open fires on the stage; use of fire-arms; carelessness in the fireworks laboratory; dancing on the stage with lighted torches or flambeaux.

Accidents to scene-hoisting machinery, also hasty shifting of light and highly inflammable scenery in too close vicinity of powerful unprotected gas-flames.

Carelessness in lighting border-lights.

Non-removal of rubbish and oily waste, and ignition of same, either by spontaneous combustion or by the careless throwing away of lighted matches.

Smoking in the theatre, in the actor's dressing-rooms, and on the stage. Careless throwing away of lighted cigars or cigarettes.

Careless handling and upsetting of candles or oil-lamps on the stage.

Storage and undue accumulation of large quantities of scenic decorations or inflammable material of any kind near the stage.

Careless handling of actors' wardrobes in dressing-rooms.

Use of stage or of loft over auditorium for a carpenters' or painters' shop.

Even where the indirect causes enumerated under (*a*) are avoided by having a theatre suitably located, well planned, and properly constructed, the more direct causes of fire mentioned under (*b*) should be carefully guarded against by good management, strict rules and regulations, by having a well-drilled staff of stage employés, and by instituting periodical inspections of all the details forming together the equipment and interior arrangement.*

Statistics of Theatre Fires.

Statistics of theatre fires show that out of a list of two hundred and eighty-nine fires known up to the year 1878,

19 per cent started in daytime.
5 " " one hour before beginning of performance.
12 " " during the performance.
24 " " within two hours after the performance was over.
39 " " during the night following the performance.

In 1881, when these statistics included three hundred and seventy-three theatre fires, the percentage remained almost the same, viz.:

19.9 per cent of the fires started in day-time.
5.6 " " " " one hour before the beginning of the performance.
11.6 " " " " during the performance.
22.6 " " " " within two hours after the performance was over.
40.3 " " " " during the night following the performance.

These figures would tend to show a large preponderance of fires immediately following a performance, and prove conclusively that the safety of theatres depends largely upon a careful and minute inspection of the building after each performance. The danger of fire, notwithstanding the above figures, is really greatest *during* a performance, but the number of casualties at

---

* See also Chapter I.

these hours is not so large as might be expected, because of the stricter and more careful watching, upon which a judicious theatre management should wisely insist.

The number of theatres annually destroyed by fire is very large. According to the carefully-collected statistics of Herr Foelsch, two hundred and nine theatres were burned in the eleven years from 1871 to 1881, making an average of nineteen such buildings per year.

From 1882 to 1888 theatre fires have occurred as follows:

In 1882, 25 fires.	In 1885, 8 fires.	In 1887, 18 fires.
" 1883, 22 "	" 1886, 8 "	" 1888, 15 "
" 1884, 10 "		

Many of these fires signify not only loss of property, but loss of life as well. The large theatre-fire calamities of Brooklyn, Nice, Vienna, Paris, Exeter (England), and Oporto (Portugal) alone were the cause of the loss of about one thousand six hundred human lives.

The large number of fires annually occurring in theatres, and the incidental great danger to the people in such buildings, is easily explained by the many possible causes of fire cited above.*

General Considerations.

If we consider the building only as such, the subject naturally divides itself into the prevention and the extinction of fire. But, looking at it from a different

---

* See also Chapter I.

point of view, and taking into consideration that during certain hours of the day or evening the building is occupied by a vast crowd in the auditorium, and by actors, actresses, supers, chorus, ballet-girls, and stage employés, often numbering several hundred people on the stage, the safety of the building becomes of secondary importance in case a fire should break out shortly before or during a performance. It then, from a humane point of view, clearly becomes chiefly a question of the safety of the spectators, the performers, and the stage-hands.

It follows, as a matter of course, that both the audience and the people on the stage are much less in danger in a well-planned, well-constructed, and properly equipped theatre than in a defective, antiquated, highly combustible structure, in which means for protection of life, means for quick egress, and appliances for quickly putting out a fire at its start are not provided, and where both life and property are exposed to undue risks.

Then again, in all buildings devoted to amusement, recreation, instruction, or worship, in which a large number of people congregate, it is very often not so much a question of danger from fire as it is of danger from a *panic*. Even where there is no real danger a panic often ensues. The lives of the theatre-goers, of the actors and stage employés, are imperilled, not only by the possibility of being burned to death, but they are, perhaps to a much greater degree, exposed to the direct effects of shock or fright, or to bodily injuries resulting from a rush, by falling or being trampled upon, or being crushed; and, finally, there is the danger of becoming asphyxiated by the thick smoke and the fire gases, or being overcome by the heat of a conflagration. A theatre building may be endangered by fire from without as well as from within, and in both cases, if the fire breaks out during a performance, the chief danger to the public arises from a sudden panic.

It obviously follows, *first*, that theatres must be so planned, constructed, equipped, and managed as not only to prevent fires, but also to prevent a panic; *second*, that there should be provided, in every theatre building, not only means for fire extinguishment, but also for protection, in case of an outbreak of fire, against the flames, the smoke, and the deadly gases of combustion, the smoke constituting the greatest danger, as more people are suffocated than burned to death in a theatre fire; and *third*, and most important, that to guard against a threatening panic, crush, or stampede, there should be ample means for personal safety and rapid egress.

The safety of the spectators and of the stage people is always of primary importance, and the safety of the building as regards its fire-resisting structural features is only second in order, except in so far as it naturally adds to the protection of the persons in the theatre. In other words, so far as the safety of human life is concerned, a combustible theatre may sometimes be less objectionable, if ample means of protection and safety-appliances are provided, and if the arrangement of stairs and exits is such as to permit the quick emptying of the theatre in from two to three minutes, than a so-called fire-proof theatre with insufficient safety appliances and lacking suitable means of egress.

The *prevention of fire* in theatres is accomplished by proper location, planning, construction, interior arrangement and equipment, by proper management, and by periodical inspections. But we should not forget that only the fulfilling of *all the requirements combined* provides perfect safety. For instance, a free site alone does not preclude theatre disasters. The Paris Opéra Comique stood on an open square, the Exeter Theatre had three fronts on streets, the Opera House at Nice and the Vienna Ring Theatre practically stood detached: but they lacked sufficient and proper exits; the stairs were not lighted, and in some cases the doors opened inward. Then, again, fire-proof building construction alone will not prevent the occurrence of theatre panics, where stairs are too narrow, passages ill-lighted or quite dark, or in case the smoke from a stage fire is drawn through a wire curtain into the auditorium by the suction of the ventilator over the auditorium gasalier. Again, it is not sufficient to build a fire-proof proscenium-wall to completely separate the stage from the auditorium, but the large opening in the proscenium-wall must be provided with an efficient fire-curtain and the other openings with self-closing fire-doors, to keep out the flames and smoke, at least long enough for the audience to make their escape. Finally, the most elaborate system and complicated mechanism of stand-pipes, sprinklers, iron curtain, stage-ventilator, etc., fail to protect a theatre audience if the exits are inadequate and if the building cannot be completely emptied in a very few minutes. What is needed, therefore, is the combination of all known elements of safety.

The *prevention of panic* in a theatre requires, first, the avoidance of all usual causes of fire, and, second, the elimination of defects in construction and plan, such as narrow and dark passages, winding staircases, crowded seats, insufficient number of aisles and exits, doors opening inward, and a badly hemmed-in site. The audience must know that it will be able to leave the building quickly and safely in case of a false alarm, or a panic caused by a fire-alarm in the neighborhood, or by the sudden accidental extinguishment of the lights in the theatre, or by the falling of a part of the scenery or decoration, or from any other disturbance of the performance.

The *means of protection* in case of *fire* include the use of the fire-proof curtain to cut off the audience from the sight of the fire and from the smoke and fire gases, the use of the stage-roof ventilator to remove quickly the smoke incident to a stage fire, and means for fighting the fire, such as a sufficient fire and water service, ample fire appliances, fire-alarm, etc.

The *means of safety* in the event of a *panic* are: for the audience, plenty of plainly marked exits, wide aisles without steps or obstructions, well-lighted and wide corridors and staircases, plenty of wide stairs with strong hand-rails on both sides, suitable fire-escapes, wide doors opening outward and preventing a jam; and for the theatre employés and the actors, isolated dressing-rooms, fire-proof stairs, a sufficient number of separate and direct exits from the stage to outdoors, and safe means of retreat from the fly-galleries and the rigging-loft.

From whatever point of view we consider the subject—whether we are concerned with the safety of the building and its usually valuable and expensive contents, or with the safety of the theatre audience and the stage people; whether we wish to prevent a panic or a conflagration—the principal requirements of a modern theatre building, and the principal safeguards, as given below, must, for absolute security, be insisted upon.

Location and Site.

The ideal site for a theatre is on a wide public square, the building having approaches and standing free and isolated on all four sides. For absolute safety this is really the only available site. It affords the best opportunity, not only for architectural effect, but also for judiciously providing and arranging numerous exits from all parts of the house. Such a site is, however, rarely attainable in our large cities, where the price of land is enormously high. If the requirement of a free site were insisted upon, it would cause the building to cost too much, and would thereby render more difficult theatrical enterprises which in this country are not subsidized by the Government, but are private commercial enterprises.

The opposite extreme, namely, a theatre building entirely surrounded or hemmed in by other buildings, is of course inadmissible, owing to the constant exposure of such a theatre to danger from outside, and on account of the increased difficulty in saving life in case of a fire or a panic within. Theatres having entrances or exits necessarily carried through other buildings are bad and dangerous. Windows or openings in the outside walls of a theatre looking towards other buildings in close contiguity to the same are also a constant menace and danger.

If a theatre building is located in a street block, it should have at least one front on the street, and there should be provided in the plan wide open courts at both sides. If the building is a corner building, located at the intersection of two thoroughfares, there should be one such court. The width of these courts should be in proportion to the number of people which the theatre is expected to accommodate. The New York Building Law requires the courts to be not less than 7 feet wide each for 1000 persons, 8 feet for 1000 to 1800 persons, and 10 feet for a theatre of a capacity above 1800 people. The court should begin on the line of the proscenium-wall, and should extend in front to the wall dividing the lobby from the auditorium. Separate corridors should continue from there through the front superstructure to the street. These corridors, according to the New York law, should not be reduced more than 3 feet in width, whereas it would seem much safer to continue them as near as possible the full width to the street. The outer opening should have iron folding gates or doors opening outward, and kept open during and at the end of each performance.

Better than a building on a corner lot is a theatre standing free on three sides, owing to the reduced danger from the adjacent buildings or surrounding stores and shops.

Plan of Building.

The theatre plan should be simple, compact, and clear, and as much as possible symmetrical as far as entrances and stairs are concerned.

To the mind of the theatre-going public the plan resolves itself into only two principal divisions—that *before* and that *behind* the curtain. This, in practice, hardly covers the requirements.

All theatres should be planned with a number of internal divisions, each of which is to be made in itself as fire-resisting as possible, to prevent the spread of fire from one part to another. There should be in every theatre, whatever its size may be, at least four principal and distinct internal divisions, each entirely separate and isolated from the rest, as follows:

(1) The auditorium;

(2) The stage;

(3) The dressing-rooms, workshops, offices, and wardrobes;

(4) The stairs, foyers, lobbies, corridors, passages, cloak-rooms, refreshment-rooms, the entrances and exits.

The auditorium comprises the various parts where the spectators are seated, namely, the parquet, or orchestra, and dress-circle, the balcony, gallery, the tier of boxes, and, above all these, the auditorium loft and roof. Each subdivision or tier in the auditorium should have separate entrances and exits. The number of tiers in all theatres, except very large opera-houses, should be restricted, because the danger is increased as the building increases in height. As a rule, there should be only two tiers—the balcony and the gallery. The fewer steps necessary to reach the street from the different parts of the auditorium, the safer will the audiences be. In nearly all large theatre catastrophes the largest number of those killed were persons who occupied the upper galleries. For the same reason it should be prohibited to place the main auditorium on the second floor for the sake of utilizing the street floor for stores. The New York Building Law requires that the main front entrance to the theatre be not more than four steps above the sidewalk. It is desirable that there should be wide corridors running around each gallery or tier, of capacity to hold the total number of persons on each division. There should be no obstructions of any kind in these corridors.

The stage may be subdivided into the under-stage, the stage proper, the wings, the fly-galleries, the rigging-loft or gridiron, and the stage roof.

The stage is the special domain of the actors and singers, of the scenic artist, of the theatre machinist, and of the gas or electrical engineer. It is also the chief point of danger in a theatre, because of the large amount of constantly changed and shifted combustible scene material, and because of the numerous and powerful illuminants and their incidental heat and liability to set decorations or hoisting-machinery on fire.

The general term " stage " includes several divisions, of which the stage floor proper is the largest and constitutes the scene of the play. The stage floor has on each end immovable sides or wings, whereas its middle portion is so framed as to be changeable or removable. As a rule, the stage floor is built with a slope towards the foot-lights, although the more recent theatres have level stage floors. The floor is constructed of narrow, well-matched, tongued-and-grooved boards of well-dried pine, and it is planed absolutely smooth. The space directly under the stage is taken up by the traps operating the movable parts of the stage floor, and by the machinery belonging thereto.

The fly-galleries are located on both sides of the stage at such a height above the wings as to enable the proper manipulation of all borders, drops, etc. As a rule, smaller theatres have only one fly-gallery, but sometimes two or more are arranged, one above the other. In some theatres the fly-galleries are connected at the back of the stage by a painters' bridge thrown horizontally across the stage, and either made stationary or else arranged so that it can be raised or lowered.

Above the fly-galleries the rigging-loft or gridiron is built across the whole stage, and is covered with narrow slats, laid some inches apart. Here are located the blocks and drums, the ropes and pulleys to which the drops, borders, and curtains are attached. Above the rigging-loft is the stage roof, with the stage ventilators or sliding skylights.

The dressing-rooms for actors and actresses are generally arranged in tiers on one or on both sides of the stage. Of dressing-rooms there is a more or less large number, according to the size, purpose, and special use of the theatre; and besides the single rooms for the " star " and the prominent actors, there should be larger dressing-rooms for the " supers," for the male and female chorus, and for the members of the *corps de ballet*. Dressing-rooms under the stage, or in the basement, generally without light and air and unsanitary, are an abomination, and should be prohibited. This division of a theatre generally also contains the theatre-manager's office. In larger theatres there should be, in addition to those mentioned, a room for the stage-manager, a room for the music conductor, a room for the orchestra, rooms for rehearsal of the chorus and ballet, offices for the scenic artist, rooms for the painter, the costumer, the property-master, the master or stage carpenter, the stage machinist, the gas engineer and the electrician, the chief engineer in charge of the boiler and pump plant, and other members of the theatre staff, besides a large room for wardrobes and costumes.

Scenery-making, scene-painting, and carpentering on a large scale should not be permitted in the theatre proper. There should be provided, in a separate annex or fire-proof building, such paint-rooms, carpenter-shops, scene-docks, store-rooms, for timber, furniture, etc., and other workshops as the business of the theatre may require. The stage, the under-stage, or the fly-galleries should never be used as a carpenter or paint shop, nor should such work be performed, as is often the case in the older theatres, in the loft over the auditorium. Neither should this loft be appropriated for the storage of inflammable material. The rear of the stage should not be used as a scene-dock.

The combination of a hotel and a theatre, or of a theatre and office-building, which is still quite common, cannot be approved from the point of view of safety, as each part endangers the other. No living or sleeping apartments should be contemplated in the plan of a theatre; no workshop, manufacturing establishment, or storehouses should be combined with it; and on the street floor stores should only be provided if separately fireproofed, if completely isolated from the theatre by thick brick walls, and if the stores are kept accessible from the street only.

The furnace or heating apparatus and the steam-boiler room should never be placed under any part of the stage, the auditorium, or the stairways. It is advisable, where a separate building for the boiler cannot be obtained, to place it in a vault under the sidewalk.

The fourth division comprises the foyers, the theatre lobby, the toilet-rooms, lavatories, and retiring-rooms for the public, the principal stairs for the various tiers, the corridors, passages, and the entrances and exits. The corridors, foyers, and lobbies should be roomy, and amply large to furnish standing-room for all persons in each division. The lobbies should be separated from the auditorium by fire-proof walls. Better still is the plan of having large open safety terraces or open colonnades on both sides of the auditorium, with stairs leading to the street, thereby preventing all possibility of a jam or crush.

There should be provided, for every section of the

house, suitable and well-arranged toilet-rooms for men and for women. If cloak-rooms or wardrobes for the public are included in the plan, they should be located near the entrances, but their arrangement must be such that persons who wait at the cloak-rooms do not interfere with the retiring audience passing through the corridors to the exits.

All corridors, passages, stairs, and aisles must be planned to be free from all obstructions, such as wardrobes, refreshment-rooms, or radiators, and so as to permit the quick emptying of the house. The crossing or meeting of two streams of people on their way out, either in the lobbies, or in the passages, or on the stairs, must be prevented.

The stage and the auditorium form two large spaces which cannot well be subdivided. All other parts of the theatre should be subdivided into as many distinct and separate fire risks as possible.

The four principal divisions of a theatre should be separated and completely isolated from each other by fire-proof walls and doors. Each division should be enclosed in fire-proof walls and made as incombustible as possible. Likewise should all horizontal subdivisions, the ceilings, roofs, and stairs, be made fireproof, in order to localize and confine a fire, so that it can burn itself out. There should be as few openings of communication in the various fire-proof divisions as possible, and where these openings are necessary, there should be means provided for closing them, in case of fire, by fire-doors opening outward. It is most important that the stage be completely separated from the auditorium by a fire-proof proscenium-wall, carried from the cellar to a height of three feet above the roof, the full width of the building. This wall should have only a few openings, namely, the large proscenium-opening, closed by a safety or fire curtain, of iron or asbestos, and not more than two other openings, at or below the stage level, closed by automatic or self-closing fire-doors without locks or bolts. There should be no communication between the stage building and the auditorium at any point above the level of the stage.

Construction of Building.

A theatre building, to be structurally safe, requires to be constructed wholly of fire-resisting materials. The natural tendency being to use the largest part of the available funds for the exterior and interior decoration of the theatre, it is advisable that the method of construction and the building materials used should be regulated by strict building laws.

Woodwork is to be avoided in the construction. Its use must be confined to the trim of the doors and windows, the usual doors, and the auditorium floor and platforms. The floors of corridors, foyers, and lobbies should be tiled. Wooden wainscoting should be avoided, and marble or tiling for stairs, lobbies and foyers and toilet-rooms, and stucco-work for the decoration of the auditorium and the boxes should take its place. Exposed ironwork should not be used anywhere, it being unsafe, except when cased with fire-proof material. Brick and terra-cotta should be used in preference to stone. The well-known principles of sound, fire-resisting construction should be applied to all parts of the theatre building. All main division walls should be strong and solidly built, well bonded together, and constructed of brick. These, as well as all staircase walls, should be run at least three feet above the roof to form fire-walls. All vertical and horizontal divisions should be fire-proof, and this refers not only to the main divisions of the plan, but likewise to all partitions. The external and internal walls must be well bonded together.

Wherever openings are provided in fire-walls for communication, they should be closed with fire-resisting doors. The doors should have stone saddles, and should be constructed either of boiler-iron, or, better, of double layers of planks of oak, lined on both sides and at the ends with tin.

The stage proper is the place where an immense amount of easily combustible material is necessarily crowded together. It should be the special aim to make this part of the building, where a fire once started is much more dangerous and much more difficult to fight than in other parts of the house, as fire-resisting as possible.

The floor of the stage proper must necessarily be constructed of wood, on account of the numerous traps; but the fixed portion of the stage floor in the wings or sides of the stage, the entire fly-galleries, and the gridiron, or rigging-loft, except the wooden floor covering the same, should be built of iron or steel beams and made fire-proof, and the pin rails of the fly-galleries should be of iron or steel. The roof of the stage should be constructed of iron, and covered with slate, tile, or other fire-proof material.

In the fitting up of the under-stage woodwork should be avoided. This is the special department of the stage carpenter, who usually clings to old-fashioned traditions and the use of antiquated apparatus. Much of the equipment of the whole stage, the stage machinery, the traps, the fly-rails, the hoisting apparatus, etc., could be made safer and more durable by substituting iron and steel for wood, by using hydraulic lifting machinery with steel wires in place of the old-fashioned hoists, sheaves, and ropes still so much in use. Suggestions tending to improve the construction of the stage were made by the "Asphaleia" Society of Vienna soon after the destruction by fire of the Ring Theatre in Vienna, where several hundred persons perished by fire and smoke. A number of European theatres, such as those at Halle in Germany, and at Buda-Pesth in Austria-Hungary, and the Royal Prussian "Schauspielhaus" in Berlin, have been fitted up in this manner with most satisfactory results. Up to the time of writing, there is in this country but one solitary instance of this advanced method of stage construction, namely, in the Auditorium Theatre, in Chicago. It is noteworthy that in the "Asphaleia" stage-setting much of the scenery and scene-setting is painted on thin sheet iron, held in light iron frames.

Wherever woodwork is necessarily employed in the construction of the stage and its accessories, it may be made fire-resisting by impregnation, or at least by the application of fire-proof paint. Even the ordinary whitewashing of woodwork helps to protect the same.

The proscenium-wall, forming the main division

between stage and auditorium, must be a fire-wall, built of brick, and the stage-opening should be arched over. The moulded frame or other finish, and all decorative features around the proscenium-opening, should consist of incombustible material.

The auditorium should likewise be fire-proof in all its parts. The main floor and the tiers should be constructed of fire-proof iron beams, and brick or tile arches. Iron columns for the support of iron girders should be encased with fire-proof materials. The fronts of galleries and balconies should be formed of fire-proof material. The ceilings of the auditorium, the balcony, and the galleries should be fire-proof. The auditorium roof should be constructed of iron, and the roof covered with slate or tile. All boxes in the auditorium, and the partitions between boxes, should consist of fire-proof material. Plaster should be used in preference to wood or *papier-maché* for the decorative treatment of the boxes.

The upper floors in the auditorium may be laid with wooden boards; but there should be no hollow spaces underneath, and the floors should be laid on sleepers deeply bedded in concrete. The ceiling of the cellar should be fireproofed with I-beams and brick arches, or other fire-proof construction, except, as stated above, directly under the middle portion of the stage.

The actors' dressing-rooms and all other rooms necessary on the stage end of the building should be fire-proof in all parts. Particular attention should be paid to the dressing-rooms and storage-rooms for wardrobes. The walls and the corridors separating the actors' dressing-rooms from the stage, and all dressing-room partitions, must be fire-proof. All doors should be of iron or other fire-resisting construction. All shelving and cupboards in all dressing-rooms, property-rooms, wardrobes, and storage-rooms should be constructed of metal or slate, or other approved fire-proof material.

Finally, all halls, passages, and foyers should have fire-proof ceilings and floors, and all staircases throughout the theatre should be constructed in a fire-proof manner.

All partitions must be built of fire-proof blocks. There should be no canvas covering of walls, and wooden wainscoting should be used sparingly and only when applied directly to solid fire-proof filling.

All windows in a theatre should be arranged to open, and none of the windows in outside walls should have fixed sashes or iron grilles.

Staircases, Entrances and Exits, and Fire-escapes.

The planning and construction of the staircases and exits of a theatre are of the greatest importance, because they form the principal means of safety for an excited or terror-stricken audience when a fire breaks out, or when there is a sudden panic from any cause. For this reason all stairs and exits must be so arranged as to disperse an audience as quickly as possible. The emptying of a well-planned theatre should not require more than two or three minutes.

All stairs in a theatre should be fire-proof, and encased in fire-walls. Open stair-well construction should be avoided as dangerous. The stairs should be from four to six feet wide, according to the maximum number of people for which they are intended. The minimum width for fifty persons is four feet, and six inches should be added in width for every additional fifty people.

All stairs should have easy and even treads. The risers should not exceed seven and one half inches, and the width of treads should be not less than ten and one half inches. Winding stairs should be prohibited, except in case the winders have treads not less than seven and one half inches wide at the narrowest part. Single or double steps should never be permitted, as they form dangerous stumbling-blocks. Long flights of stairs must be broken by several landings, at least four feet wide. The Boston Building Law calls for stair flights to have not less than three, nor more than fifteen, steps.

There should be, on each side of all stairs and landings, well-fastened hand-rails turned into the wall at the ends of stairs. Very wide stairs are preferably divided by a strong metal baluster or intermediate central rail. All staircase halls should be made as smoke-proof as possible. The stairs for the balcony and gallery should not communicate with the cellar, so that smoke from a cellar fire may not rise up through the staircase.

There should be at least two entirely separate stairs for each part of the audience, one for each side of the building. Inasmuch as the occupants of the gallery have the longest way to the street, and also because the experience of theatre catastrophes teaches that this part of the audience is the one most endangered in case of a panic, there should really be more stairs and exits for the highest tier, although in the theatres erected in the past this precaution has seldom, if ever, been observed. The stairs serving different tiers should not communicate with each other, and the meeting of two streams of people on the same stairs should always be avoided. Each portion of the audience, the parquet, the balcony, and the gallery, should leave by separate stairs and exits. Each tier should also have separate wide corridors running all around, and leading to a foyer and to the staircases.

It is usual to calculate two exits for 300 persons, and three for 500 persons, not including the outside fire-escapes. The more exits there are, the less liability of fierce struggles and jams during a panic, and the less danger there is of old or weak persons becoming hurt.

The planning of the means of egress, the exits, and fire-escapes in such a way as to effect a quiet and quick emptying of a theatre requires the most careful study. All aisles, corridors, passages, and stairs should be of even or increasing width, must not be tortuous, and should have no obstructions of any kind, such as pay-boxes, cloak-rooms, refreshment-counters, etc. All passages leading to exits must be at least four feet wide.

The best arrangement and judicious location of the exits depends to a great extent upon the site of the theatre. If the same permits placing plenty of wide and easily reached exits on all sides of the building, the safety of the audience is assured. The more the audience can be subdivided, the quicker can a theatre be emptied.

The minimum width of exits for 500 or less persons, according to both the Boston and New York theatre laws, is five feet, and twenty inches shall be added in width for each additional 100 persons. The exits must likewise be free from projections or obstructions of any kind.

Many lives have been lost in theatre catastrophes owing to faulty arrangement of the doors. It is imperative that all exit doors be made to open outward, to prevent a jam in case of a panic. The doors in corridors must, moreover, be so hung as not to become an obstruction in the passageway. Sliding doors should never be used. Emergency exits or doors, which are frequently advocated, are not desirable. They have often been found locked when needed, and, moreover, the people are not familiar with their location. *All* stairs and exit doors should be opened for the use of the public at each daily and nightly performance. Exit doors should have proper bolts at shoulder height.

All exits should be as conspicuous as possible and should be plainly marked in large legible letters, and all doors not leading to exits should be lettered " no exit."

In addition to the separate stairs mentioned above, each tier should have exits to at least two well-constructed outside iron fire-escapes, to be used in case the retreat by the inside stairs is cut off by smoke. All such fire-escapes, to be of any use, must be so constructed with regular steps that they can be used by women and children. These outside fire-escapes should be covered with metal hoods or awnings to prevent, as much as possible, their becoming slippery and difficult to use when covered with snow and ice in winter weather. The New York Law requires the fire-escape balconies to be four feet wide, and the stairs to have not more than eight-and-one-half-inch risers and at least nine-inch treads. They must be built sufficiently strong to carry the weight of a large crowd.

The same care and attention is required in the construction of the stairs for that part of the building containing the stage, the offices, and the dressing-rooms. The stairs and exits for the actors and stage hands should be entirely separate from those for the spectators, and there should be at least two such stairs and exits, preferably one at each end of the stage. There should also be fire-proof stairs leading from the fly-galleries and the rigging-loft, one on each side of the stage, these places being, in case of a stage fire, among the most dangerous, owing to the rapid rising of the smoke and the flames. It will be remembered that at the burning of the Paris Opéra Comique the majority of the audience escaped, while many chorus-singers and ballet-girls were burned to death, because the stage exits and stairs were lamentably insufficient.

Finally, there should be well-fastened iron ladders provided on the outside courts and leading to all roofs of the building, for the use of the firemen.

Aisles and Chairs.

The width and number of aisles in the parquet, balcony, and gallery depend upon the size of the theatre and the number of seats provided. Many narrow aisles breaking the continuity of the seats are better than a few wide aisles. Aisles should be straight and leading directly to the exits, or else the direction of the nearest exit should be indicated on the wall by a conspicuous arrow. The aisles should widen with the number of rows of seats which they serve. The minimum width of aisles is three feet, to which should be added one and one-half inches for every five feet length of aisle.

In the orchestra, or parquet, there should be no steps in the aisles, but the aisles should be inclined under a gradient of not more than one in ten. Gradients should likewise be used for differences in level between the aisles and the open courts.

Aisles should never be obstructed by camp-stools or movable chairs, which become a source of danger when overturned during a panic, nor should standing in the aisles be permitted.

The width and depth of the chairs or seats should not reach below a certain minimum, both for comfort's and safety's sake, and there should be not more than thirteen seats between two aisles, so that no seat in the audience has more than six seats between it and the aisle.

The framework of the chairs should be constructed of fire-proof materials, the chairs should be firmly and well bolted, or otherwise fastened to the floor, and the seats should preferably be automatic, i.e., self-raising. In any case they must be hinged so as to be easily raised and placed out of the way to facilitate the emptying of the rows of seats.

Movable chairs should be allowed only in the boxes,

and for each of these the number of seats should be limited.

Fire-proof Curtain.

The stage opening in the proscenium-wall should be closed by means of a fire-proof curtain, entirely separating the stage from the auditorium. The object of the curtain is to prevent a fire on the stage or in the flies from leaping over into the auditorium, and to interpose a barrier to the smoke and fire gases generated in a conflagration. It is, therefore, one of the chief means of safety for the audience, and in theatres where it is installed, kept in good working order, and let down at the first moment of danger it helps to enable the audience to escape unhurt from the burning building.

Numerous suggestions have been made regarding the material and construction of fire-proof curtains. The three principal forms adopted in practice are a curtain of wire gauze, a curtain built of corrugated iron, and asbestos woven curtains. In European theatres the first two kinds have been tried, whereas the asbestos fire-proof curtain is chiefly used in American theatres.

Curtains made of wire gauze with fine mesh have not proved successful in actual use, because while they keep the flames from spreading across the stage opening into the auditorium, they allow the smoke to pass, and also because they increase the panic and confusion by giving the frightened audience a view of the fire on the stage.

Corrugated-iron curtains are better in this respect. They should be accurately guided at both ends and should travel easily in proper vertical metal grooves placed on both sides of the proscenium-opening, and well fastened to the brick wall, otherwise they are liable to stick fast at the moment when wanted. Such iron sliding-curtains have not, in practice, proved themselves always reliable. It is important that the apparatus for raising or lowering the fire-curtain should be on the stage, and not in the rigging-loft, as was the case in the Vienna Ring Theatre, because this point may be beyond reach soon after the outbreak of a fire on the stage. Unless built very strong, iron curtains are liable to buckle and warp, and do not resist a strong pressure due to the expansion of the air by the heat in case of fire. They should fit tightly, to prevent the escape of smoke, and should be strong enough to sustain a pressure for at least ten minutes, to give the audience time to escape. If they are not protected from above by automatic sprinkling apparatus, or a series of perforated pipes connected with the supply-pipes or roof-tanks, so as to allow a sheet of water to run down along the curtain, they may become red-hot during a fierce fire and thus endanger the spectators.

On the whole, thick woven asbestos fire-proof curtains are lighter and more easily handled than iron curtains, and prove well adapted to check the spread of flames and to keep back the smoke, at least sufficiently long to permit the complete emptying of the theatre. The asbestos curtain should travel in grooves and should fit the proscenium-opening as closely as possible.

Whatever kind of fire-proof curtain is chosen, it

should not merely be used in case of an emergency, but nightly, and it should be raised a few minutes before the beginning of, and lowered immediately after, the performance.

It is a good plan, as a further protection of the proscenium-opening, to provide it with a 2½-inch perforated copper pipe, fed from the sprinkler tank by means of shut-off valves operated from the prompter's side of the stage, the descending stream of water forming, when set in operation, an efficient water-curtain.

Stage Ventilator.

Every theatre stage should be provided in its roof with one or more large metal flues or ventilators, or with ventilating sliding skylights, operated by means of ropes and counterweights controlled at the level of the stage, or arranged to work automatically by the burning of a hemp cord in case of a stage fire. The object of this ventilator is to provide an outlet and ready means of escape for the thick smoke and the fire-gases. Such a ventilator would also act as a means of increasing the draught and spread of the fire, and on this account it has been objected to by many who were evidently more concerned with the saving of the building from destruction than with the saving of life. But, human life being of greater value than property, the objection raised is evidently of little importance, considering the fact that the ventilator acts as a powerful means of removing the smoke. As frequent theatre catastrophes have sufficiently demonstrated, the smoke is a more deadly agent than fire, and constitutes the chief danger to the audience. Incredible as it may sound, it is nevertheless true, that, on the average, only five minutes elapse between the appearance of fire in front of the proscenium-opening and the total extinction of human life by smoke and fire gases. This worst of all foes, smoke, must be kept from the audience by all known means.

The two appliances, a fire-proof curtain and a stage ventilator, act combined as the most important means of protection to the audience in case of a theatre fire. Where these are installed it is, on the other hand, absolutely necessary to provide safe means of retreat for the stage-hands employed in the rigging-loft and in the fly-galleries.

In Boston the Building Law requires that the combined area of the stage ventilators be equal to one-tenth the area of the stage floor, and each ventilator is to have a counterbalanced valve, so as to open automatically, the valve being kept closed ordinarily by a cord of combustible material, run down to the prompter's desk.

The New York Building Law requires metal sliding skylights in the stage roof, of a combined area equal to at least one-eighth the area of the stage. These skylights are to be fitted with sliding sashes, glazed with double-thick glass, not exceeding one-eighth inch thick. Each pane is to measure at least 300 square inches. The skylights are to be so constructed as to open instantly on the burning or cutting of a hempen cord, ordinarily arranged to keep the skylight closed.

A very ingenious apparatus for stage ventilation was fitted up some years ago in a Chicago theatre. The proscenium-wall was made fire-proof, and all openings in it were stopped by fire-resisting doors provided with springs to keep them closed. Over the stage an exit for smoke is provided, consisting of a boiler-iron flue, 8 feet in diameter and 30 feet high. In this flue is a valve made of a wooden frame covered with canvas, which is kept closed by a balance weight. A wire-cable connection is carried down from the valve to each side of the stage, and there are other cables carried to points in the auditorium under control of the ushers. By pulling any of these cables the valve is opened, while a very simple and ingenious contrivance closes simultaneously the ventilating outlet over the audience chandelier, thus reversing the usual movement of air from the stage to the auditorium ventilator. Should the cable not be worked in time, the valve being of canvas is readily destroyed by fire, and the smoke outlet in the stage roof is thereby opened. The same contrivance may at ordinary times be used for the ventilation of the stage, and during performances it should be opened when it is desired to remove the powder smoke incident to fireworks or the firing of guns.

Fire-proof Treatment of Stage Scenery.

The usual stage scenery and scenic paraphernalia constitute a conglomeration of highly dried-up wood-work, a perfect labyrinth of ropes, and a vast quantity of gauze borders and hanging drapery. This mass of highly combustible and easily inflammable material forms a constant danger and menace to the stage, particularly when brought in close vicinity to gaslights.

It should be the aim to make the various parts of the stage scenery, such as the wings, the borders, the drop-scenes, and the set-pieces, as fire-resisting as possible. The substitution of light iron framework in place of wood for attaching the painted canvas is desirable. Where wooden frames are used they should be coated with fire-proof paint.

All scenery, draperies, and furniture should be rendered non-inflammable by chemical treatment. The treatment, to remain effective, should be renewed periodically. This is easily accomplished where a theatre is used by the same company during the entire season, but is difficult to insist upon where different travelling companies occupy the theatre in succession. The fire-proof treatment of scenery is required in the New York theatre law, but practically the law is a dead letter.

The woodwork constituting the floor of the stage proper and the traps in the under-stage should be rendered uninflammable by chemical treatment.

All scenery for large theatres should be stored in a separate fire-proof scene-dock, forming an annex to the theatre building.

Stage Construction and Stage Machinery.

I have already stated that the stage construction and machinery of the majority of theatres is antiquated, and that the system followed from time immemorial is susceptible of much improvement. In fact, what is wanted is a radical improvement and innovation, tending to render this part of the building much more secure, and at the same time more convenient in management.

In a paper in the *Journal of the Society of Arts*, on "Scenic Illusions and Stage Appliances," Mr. Percy Fitzgerald, F.S.A., relates that "for the new Paris Opera-House, built by Garnier, a novel plan of stage construction was offered and seriously entertained. This was to divide the whole stage into small platforms, each supported on pistons moving up and down in hydraulic presses. A lever, put in motion by the stage-manager, would thus elevate or depress any section of the stage to the height or depth required." He continues by saying: "This was ingenious, and it was elaborated with care and nearly adopted, but the objections were insuperable. The space below the stage was lost, being filled with pumps and apparatus; there were nearly a hundred pistons, but the real danger was the almost certainty of some part of the machinery getting out of order." He adds that "the system was actually adopted at the new Vaudeville, but never came into use."

The system rejected for the Paris Opera-House has since then been introduced into a number of smaller theatres, with absolute success. Its great advantage consists in reducing the risk of fire, by substituting simple iron construction, iron supports, and iron machinery for the innumerable wooden traps under the stage, and by replacing the inflammable canvas and gauze scenery stretched on wooden frames by scenic decorations painted on sheet iron, held in light iron frames. Woodwork is used only for the stage platforms, but not for their support. This is the new system of the "Asphaleia" Society of Vienna. It has been used on the stages of the new theatres at Buda-Pesth and at Halle, and also partially in a Berlin theatre. Independent of this system, several other theatres have adopted iron in place of wood for construction, and steel wires in place of hemp ropes for hoisting apparatus; for instance, the new theatre in Rouen, France; the new Hofburg Theatre in Vienna, and the Royal Theatre in Edinburgh.

The Buda-Pesth Theatre is considered to be one of the most scientifically and perfectly equipped theatres in existence. Its whole stage floor is divided into many traps of different size and shape, some circular and some oblong, which are operated by pistons worked by hydraulic power. By touching a lever any section can be raised, lowered, or inclined, and thus the stage may be transformed into terraces or gradual inclines, instead of the usual method, consisting in building up such scenery on the stage. The rigging-loft is constructed entirely in iron. The scenery of the "Asphaleia" system is equally novel. There are no flat drop-scenes and no side-scenes or *coulisses*, and borders are done away with. The whole stage forms a clean open space. In this is hung from above a large continuous curtain, or "horizon," as it is called, open toward the front and forming three sides. This is painted so as to represent different sky effects, and the horizon may be moved, and by the use of proper light the scene is readily changed. The movements of the horizon are also controlled under the stage by means of hydraulic machinery. The scenes are represented by detached pieces.

Apart from the novelty of the scenic effect, this new system is highly commendable, because it eliminates the chief elements of danger from fire in theatres, and also because it replaces hand labor in the shifting of scenery and rearrangement of the stage by a scientifically constructed mechanical power-system.

Heating.

The heating of a theatre should always be done by a centralized system, in order to avoid the necessity of having many fires to look after, and of having a number of smoke-flues, each of which constitutes a danger. Heating by warm-air furnaces is only adapted for very small theatres, and even in these it is objectionable on account of danger from fire, and because it rapidly desiccates woodwork and scenery, and renders the same still more easily inflammable. The choice of the system for larger buildings lies between warming by steam and by hot water. Whichever system may be preferred, it is essential that the heating of the auditorium be accomplished largely by the indirect method, which consists in the introduction of a constant supply of fresh air suitably warmed at heating stacks, placed at the basement or cellar ceiling. The indirect-heating system may be supplemented by a few direct steam-radiators placed in wall niches near the exits, where there is frequently an unpleasant draught from the rush of cold air from outdoors.

The stage is generally heated by direct-heating coils or radiators, and care should be taken to keep the stage building at such a temperature, even when there are no performances, as to prevent the freezing of the water in the fire stand-pipes, and in the automatic-sprinkler system.

The dressing-rooms, toilet-rooms, offices, stairs, corridors, foyers, and lobbies are also heated by means of direct radiators.

The steam-boiler must never be placed directly under the auditorium or the stage. It is desirable to place it in a separate and detached building, or, at least, to remove it to a vault under the sidewalk. The boiler should be of the sectional safety type. Special care is, of course, required in the construction of the boiler-flue. The boiler-room should be enclosed by brick walls, the ceiling should be fireproofed, and all doorways leading to the boiler-room should be of iron, or, better, double tin-lined wooden doors. If the boiler-room is equipped with automatic sprinklers, care should be taken to have the fusible solder-joint arranged so as to stand a higher temperature without opening. All steam-pipes must be kept from direct contact with any woodwork, and they should be encased with double thimbles where they pass through floors, to avoid the danger of wood becoming charred and ignited.

Floor registers are objectionable, and are not permitted by the New York theatre law. Heating-coils or radiators should not be placed in any aisles, passages, or corridors where they would constitute an obstruction, but should be placed in recesses or niches.

Lighting.

The lighting arrangements of theatres, and particularly of the stage and the stage scenery, require the most thoughtful consideration and the greatest care in execution and arrangement, as they form one of the chief causes of theatre fires. The days when theatres were illuminated by candles or oil-lamps are happily past. Gas-light, which is infinitely safer than either method, has long ago taken their place, and is now as rapidly being replaced by the still safer, and therefore much to be preferred, incandescent electric light.

In the auditorium gas-lights have proved objectionable on account of the immense heat and vitiated atmosphere which they create. On the stage, all open flames used in the lighting up of the wings and borders, as well as the footlights, were a constant menace to the highly inflammable scenery and to the light costumes of the dancers. It is therefore safe to assert that, at least for the two principal parts of a theatre building, electric-lighting has many advantages over gas-light. The introduction of the incandescent glow-lamp for the lighting of the stage and scenery increases the safety of a theatre-building perhaps more than any other safety appliance. Of course it is necessary that the wiring should be done in the best manner, according to the rules of the fire underwriters.

But the electric current is not always available, and the installation of a special electric plant may, in some cases, particularly for smaller theatre enterprises, prove too great an expense. Therefore, where the theatre depends on the use of gas, the gas-piping should be arranged with extreme care, and the management of all gas-lights should be under the control of a specially trained employé. All gas-pipes should be of wrought iron, and no lead or other pipes should be permitted. The piping must be put together in a proper and workmanlike manner, and must be absolutely gas-tight under a severe pressure test. The joints of flexible india-rubber gas-pipe for the border and bunch lights and for stage chandeliers must be carefully watched, and the rubber tubing should be strong steam hose of best quality rubber, and protected from injury by being surrounded with spiral wire.

The gas-meters should be placed in well-ventilated brick vaults, closed with fire-proof doors. There should be at least two distinct supply mains, one for the stage and the main lights in the auditorium, and another for the staircases, corridors, lobbies, the entrances and exits, and the rear portion of the auditorium. This second line may also be used for supplying the dressing-rooms and offices in the stage building, although a third separate gas-main for these would be still better. Each main should be controlled by a shut-off valve, located in the sidewalk and accessible from outdoors, to turn off the supply of gas if not required. The stage gas-main should lead to a large gas-table, or distributor, located on the prompter's side of the stage, and at this table there should be a series of shut-offs, each with by-pass to control, separately and independently, the lights in the auditorium, the central chandelier, the footlights, the wing-lights, the border-lights, the ground-lights, the stage chandelier, and the bunch-lights. The lights in the stairs, passages, lobbies, should be controlled only by a shut-off located in the lobby.

All gas and electric lights surrounded with glass should have fastened underneath a fine-mesh wire netting. All open gas-flames must be protected by wire cages or guards, which should be strongly fastened to the gas-brackets. The wire cages should be of sufficiently large diameter (at least ten inches) so that the wire may never become heated to more than 250° F. All side-lights should have stiff brackets. Swinging or jointed brackets should not be permitted. All fixtures should have strong pin stops. It is of advantage to have all gas-brackets in foyers, halls, and stairs so arranged as to have detachable keys, so that the lights cannot under any circumstances be tampered with. All these lights should be protected against draught by glass globes, but on the stage and in the dressing-rooms all open flames should have wire cages.

All open flames should also be kept safely away from woodwork. The footlights should have a wire network, and should be protected with a strong wire guard, two feet distant from the footlights, to prevent actors from coming too near to the lights. The trough for the footlights should be formed of fire-proof material. All border-lights should be suspended by wire ropes. The lights in the wings should not come lower than five feet above the stage level.

The central chandelier in the auditorium, the footlights, and the border-lights, as well as all other rows of gas-lights, should be lighted with electric flash-lighting, which is a safer method than the lighting up by means of spirit-lamps on poles, or by other open lights. Sun-burners in the auditorium ceiling are preferable to chandeliers, and are usually arranged so as to assist in the ventilation of the main body of the theatre. The central lustre is hurtful to the eyes and in the way of the sight-line of the upper gallery, and its entire abolishment is much to be desired. All ducts used for carrying heated air from gas-lights should be made double, and constructed of metal with intervening spaces. The ceiling ventilators in the auditorium should have valves arranged in such a way that they may be shut off to prevent smoke in case of a fire being rapidly drawn up towards and through them, before the audience has departed.

Provision should be made for plenty of gas-light in all actors' dressing-rooms, otherwise the actors may feel tempted to use candles in " making up."

Where lime or calcium light is used on the stage for special illuminations it should be most carefully handled and manipulated.

Every theatre should, in addition to the gas or electric light, have for the sake of safety an auxiliary system of lighting the passages, corridors, and all stairways and exits by means of oil-lamps or candle lanterns, so that in case the gas is turned off, as happened in the Vienna Ring Theatre and the Brooklyn Theatre fires, or in case the electric-light fails, the exits and stairways are not left in total darkness. These oil or candle lanterns should be placed out of the way, in niches in the walls, closed by glass doors, so they cannot be extinguished by a draught of air or by smoke. It is advisable to provide the lamp niches with fresh-air supply and a vent-flue carried outdoors. The oil burned in these lamps, as well as in those used on the stage in the presentation of plays, should be vegetable or colza oil, or sperm or whale oil, and not the mineral oil or kerosene. All oil should be stored in fire-proof vaults, and the trimming and filling of the lamps should only be permitted in daytime. The oil lamps at exits should be provided with red and white glass, so as to be plainly distinguished. The auxiliary lights should be kept lighted from the time the theatre is open until the theatre and stage are emptied. The New York law requires all parts of the building devoted to the public, all outlets leading to the street, including the open courts and corridors, to be lighted during each performance, and to be left lighted until the whole audience has departed. It also specifies the use of at least two or more oil-lamps on each side of the auditorium in each tier, the same to be placed on fixed brackets about seven feet high, to be filled with whale or lard oil, or else candles.

Lightning-rods.

I have not been able to find on record a single case of the burning or destruction of a theatre from being set on fire by lightning,* but there are instances of

---

* Soon after writing the above, I received information from Mr. Ernest E. A. Woodrow, of the burning of the Royal County Theatre at Reading, England, on the last Saturday in August, 1894, in the morning, which fire was said to be caused by lightning. I quote from a copy of the Reading *Observer* of Sept. 1, 1894, the fol-

theatre buildings having been struck. On March 20, 1784, lightning struck the theatre at Mantua, in Italy, during a performance, at which about 400 spectators were present. Of these, two were killed and ten severely injured. On the 26th of July, 1759, the theatre at Feltre, in Italy, was struck by lightning, and Arago relates that many in the audience were killed or injured, and that the lights were extinguished by the electrical shock. In both cases the building did not take fire. There is always, in the event of a theatre being struck by lightning during a performance, the danger of a serious panic. It, therefore, seems proper and wise to provide theatre buildings with the protection which a well-constructed system of lightning-rods affords. The Frankfort-on-Main Opera-House is protected by thirteen lightning-rods, subjected to inspection annually.

Ventilation and Sanitation.

The efficient ventilation of the auditorium, of the stage, of the actors' and supers' dressing-rooms, and of the toilet-rooms; a thorough system of cleanliness and removal of dirt and rubbish; and, finally, a safe and well-arranged system of drainage and plumbing—are important requirements, constituting together the question of "Theatre Hygiene," but they have very little direct bearing on the subject of fire-prevention

---

lowing: "There is little doubt but that the origin of the fire was due to the building being struck by lightning. . . . The lightning struck the top of the theatre near the ventilator over the gallery, and almost immediately the building was in flames." The theatre was completely wrecked by the fire.

and fire-protection in theatres, and will, therefore, not be discussed here in detail.

The Boston Building Law wisely prescribes that every theatre should have a system of ventilation so contrived as to provide fifty cubic feet of fresh air per minute, or three thousand cubic feet per hour, for each occupant, and for each light other than electric light.

The old-fashioned central chandelier, or lustre, which constituted a chief feature in the older theatres, and which when gas-light came into use was utilized to assist in the ventilation of the auditorium, is happily going out of fashion. In more than one instance it carried death and destruction, by creating during a fire a strong current of air from the stage into the auditorium, thus leading the smoke and fire gases to the upper galleries and suffocating the people. Serious accidents have also occurred through the falling of the chandelier during a performance.

I do not wish to be understood as being opposed to ceiling ventilation in the auditorium. This is necessary, even in the case of theatres lighted with electricity. But a judicious arrangement of the ventilators and registers is required. The ventilator over the stage, when fully opened, should be larger than the combined area of the ventilators in the auditorium ceiling. This, together with the fact that the stage ventilator is nearly always higher than the one for the audience, would assure a movement of air away from the auditorium towards the stage. To make quite sure of this, it is a good plan to have the vent registers in the ceiling of the auditorium controlled from the stage, so that in case a fire breaks out on the stage, the person whose duty it is to lower the fire-proof curtain, and to open the stage-roof ventilators, will also simultaneously close the vent register in the auditorium ceiling. The smoke and fire gases will thereby be effectively prevented from passing over into the auditorium. In any case, the ventilation of the stage and the auditorium should be arranged entirely separate.

Fire-service and Fire-extinguishing Appliances.

A theatre should have an abundant and never-failing supply of water for fire-extinguishing purposes, and there should be plenty of fire-extinguishing appliances always kept in readiness and free from encumbrances.

First, as regards the main in front of the theatre: this should be, if possible, an eight or ten inch main, and where the city water system has different pressure zones, the main should be connected with the high-pressure service.

There should be, for outside protection, several large post hydrants distributed around the sides of the theatres.

The size of the service main supplying the theatre should be at least six inches, if attainable, and four inches should be the minimum size. It is still better to run two mains into the building from opposite streets and to cross-connect the same in the building so as to be sure of an ample supply at all times.

Where water-meters are required on these services, there should be a sealed by-pass gate-valve, to be opened only in case of fire, as it is unnecessary to meter the water used for extinguishing the flames.

There are some cities in which the pressure in the city mains is at all times ample for fire purposes, and where the stream from fire-nozzles attached to stand-pipes, fed from direct pressure, would reach above the highest part of a theatre building. In other cities, where the Holly system of water-works is arranged, the ordinary pressure of about 35 to 40 lbs. in the street mains is increased in case of a fire, as soon as the fire-alarm has reached the pumping-station, to 100 lbs. pressure.

In the majority of cases, however, the pressure in the street mains, suitable for domestic purposes, is insufficient for fire protection, and it becomes necessary to supply the fire-valves and the sprinklers in the theatre either from fire-pumps, or from elevated open tanks, or, finally, from closed tanks with compressed air. Even the pressure obtained from elevated tanks is not sufficient to produce effective fire-streams, except in the lower parts of the theatre.

It is usual in theatres to operate the sprinkler system from roof tanks, and the fire-valves under pressure from a large fire-pump, and to omit any connection between the fire stand-pipes and the roof tanks.

Every theatre should, therefore, be fitted up with one or several powerful fire-pumps, either of the rotary type or of the direct-acting plunger type. The best type of direct-acting pump, the "Underwriter" standard fire-pump, should be selected. For small theatres the pump should have a capacity of 500 gallons per minute, equivalent to two fire-streams; medium-sized theatres require a four-stream pump, which is able to throw 1000 gallons per minute, while large theatres should have either the six-stream pump or two fire-pumps of smaller size. These pumps should ordinarily draw their supply from a large suction or reserve tank located in the basement of the theatre, and supplied from the street mains, but they should be so connected as to be able to draw directly from the street main in case of fire. The fire-pump should also be used to feed the open roof tanks for the sprinkler system. This pump should be kept in working order, and tested at least once a week. During performances, steam under high pressure (at least 50 lbs. per sq. in.) should always be kept on the pump to have it ready for instant service.

There should be numerous fire-stand-pipes distributed throughout the theatre, at least one on each side of the proscenium-opening on the stage, at least one on each side in the auditorium, one in the corridor near the stage dressing-rooms, one in the property-room, one in the carpenter's shop, and others as may be required. Each stand-pipe should have outlets in each tier of the auditorium, and likewise in the understage, on the stage, in the fly-galleries, and in the rigging-loft. Each outlet should be provided with a standard fire-valve, and with 50 feet of fire-hose with hose-coupling and fire-nozzle always attached to the valve.

The hose should be of a suitable and approved quality, either rubber-lined cotton or unlined linen hose, and both kinds should be able to stand a pressure of several hundred pounds per square inch without leaking or bursting, and should be tested before acceptance. Some of the unlined hose sold in the market leaks like a sieve, and even much of the rubber-lined hose as used for inside fire protection is of poor quality and utterly unfit for use. I saw the story related, though I cannot now recall where, that a dealer in fire-hose once asked a purchaser whether he wanted the hose for actual use, or merely "to hang up" to satisfy the Underwriters' inspectors. Whether the story be true or not, it points out a moral and lesson which those who fit up theatres should bear in mind when purchasing fire-hose.

The hose is ordinarily kept on hose-reels, but it is better to fold it up in swinging hose-racks, or to hang it up on saddles, hooks, or pegs in such a way that if taken out for service it does not twist or kink. There should be at each fire-valve suitable hose spanners. All stand-pipes should be kept clear from obstruction, and the access to them should not be blocked. The fire-pump should be fitted with an automatic regulator, so that immediately when a fire-valve is opened the pump is started in action. An efficient substitute for fire-hose and play-pipes consists in the use of so-called "Monitor" nozzles, attached directly to the stand-pipes, which can be turned vertically as well as horizontally in all directions, and are so balanced as to remain in position while the stream plays. These special fire-nozzles are applicable in particular to the protection of the theatre stage.

A separate and distinct system of automatic sprinklers, with fusible plugs, supplied from a fire-tank on the roof, and not connected in any manner with the stand-pipes, should be provided. The sprinklers should be installed above-and around the proscenium-opening, on the ceiling or roof of the stage, under the rigging-loft, under the fly-galleries, and under the stage, at such intervals as will protect every square foot of stage. Sprinklers should also be installed in the carpenter and paint shops, in the store-room and property-room, and in the boiler-room. It is desirable to provide sprinklers in all dressing-rooms, and where the loft over the auditorium is used for any purpose this should also be so protected.

The sprinkler system should be provided with one or several outside fire-department connections to which the fire-engines may be hitched, and with automatic inside and outside fire-alarm. The fire-tank should have a low-water alarm that will ring a bell in the pump-room.

Sometimes the stage is protected by a perforated-pipe system in place of the automatic-sprinkler system, and the valves controlling the flow of water to the rows of pipes are operated from the same place on the stage where the fire-curtain, the stage ventilator, and the auditorium vent registers are operated.

In addition to the fire-valves and the sprinklers, each theatre should have a number of portable chemical or compressed-air hand fire-extinguishers, and a large number of fire-pails or buckets. The fire-pails should be hung on hooks or set on shelves, and placed on the stage, in the fly-galleries, in the rigging-loft, under the stage, and on all the tiers of the auditorium, on both sides of the theatre. Fire-pails should be painted a bright red and marked " FOR FIRE." They should be constantly refilled, and must not be used for other purposes. Sometimes so-called "insurance" or "chemical" fire-pails are used, which contain a non-freezing, fire-extinguishing solution, in liquid or in powder form, and which are protected by tin-foil against evaporation.

On the stage there should also be large casks of water, of from forty to fifty gallons capacity, to refill the pails and to form a supply for hand force-pumps. Moreover, there should be provided wet sponges or swabs on long poles, wet blankets, asbestos sheets, asbestos gloves, boxes of sand to extinguish burning oil, various lengths of fire-hook poles, etc.

Finally, there should be on each floor of the auditorium, on each tier of the stage, and in the corridors dividing the stage proper and the dressing-rooms, fire-axes with pick heads, axe-brackets, and knives for the use of the firemen and the theatre employés.

Steam-jets or nozzles for extinguishing fires may be fitted up in inaccessible corners and in the rigging-loft, but neither on the stage nor in the auditorium should steam be used as an extinguisher, as it would tend to increase the fright and confusion during a fire or panic.*

Life-saving Appliances.

When a fire breaks out in a crowded theatre the first thought should be the safety of the audience and of the people in the stage-house, and the saving of property or of the building should be a second consideration. The fire-proof curtain and the stage

---

* See Chapter III. for further details.

ventilator may be classed among the life-saving appliances. We may also consider as arrangements tending to save life, the provision of plenty of wide aisles, and of numerous fire-proof stairs and exits. The auxiliary lighting of the latter may also be considered a necessary feature for the saving of the audience.

Special life-saving appliances, such as scaling-ladders, extension-ladders, a large jumping-net, a chute or flume-escape, rope or pulley escapes, smoke-respirators, are generally carried by the fire-brigade. Some of these—for instance, the jumping-net—may with advantage be kept in readiness at each theatre, to assist in saving life.

Fire-alarms.

Every theatre must be fitted up with a complete fire-alarm system. This should include telephonic and telegraphic connection with the nearest fire-engine station, and also with the headquarters of the fire-department, so that in case of an outbreak of fire the firemen may be immediately notified. In large theatres it is desirable to have other means to indicate quickly the presence of flames. Thermostats, or heat-indicators, may be suitably placed and distributed throughout the house, which indicate in the manager's office the presence of undue heat or fire at any point. The automatic-sprinkler system should always be fitted up with a fire-alarm, which rings a gong on the stage or in the office, and another gong on the outside of the building, as soon as any sprinkler has been put in operation. There should be a fire-alarm telegraph, connecting the manager's office with the pump or boiler-room, so that the engineer can be immediately notified, in case of fire, to look after the boilers and the fire-pump. In some recent European theatres all the doors leading to the stage and the fire-doors in the proscenium-wall are fitted with a system of alarms, indicating in the office when any of the doors are not closed. Such alarms must, of course, be switched off during the rehearsals and performances, but at all other hours they indicate whether the stage is completely shut off from the rest of the building.

Questions of Management.

There are many points in the management of a theatre which tend, directly or indirectly, to increase the safety of such buildings from the danger of fire, and which likewise affect the safety of the theatre-going public. To some of these I shall make brief reference.

Strict discipline and order should prevail on the stage. The theatre staff should be called together regularly for fire-drills, and for instruction in the use of the fire-extinguishing and life-saving appliances. Each employé should be entrusted to perform a particular duty in case of a fire.

There should be at all times a theatre watchman, and at night, in particular, a special night-watchman. His duty should consist in making frequent trips, at regular intervals, to all parts of the building, and it is advisable to have his faithfulness controlled and his inspections recorded by an electric watchman's clock.

No open lights or open fires should be allowed in any room near the stage or on the latter. The rooms for the storage of theatrical costumes should, at night, be entered only with safety lanterns. The use of candles or oil-lamps in the dressing-rooms, the use of wax or parlor matches, and likewise the smoking of cigars, cigarettes, or pipes should be prohibited. Smoking on the stage, if required in the scenes of a performance, should be restricted as much as possible. Candelabras with candles and oil-lamps, if used in the play, should be handled with extreme care.

The practice of a sudden and unannounced darkening of the auditorium, during changes in stage-setting with raised curtain, is dangerous, and may precipitate a panic.

Special care should be observed and strict regulations issued regarding the use of firearms, the burning of fireworks, the use of colored and calcium lights, so frequently introduced in spectacular plays, the dancing with lighted torches, or even the representation of actual fire scenes on the stage. The wads for rifle or pistol shots should be of calf's hair or asbestos wool, and not of paper.

Trained firemen should be in attendance on the stage during every performance. The firemen should be in charge of the appliances for the extinction of fires, and should satisfy themselves by personal inspection shortly before each performance that they are in good working order and ready for instant use. The fire stand-pipes and valves should not be obstructed. There should be a penalty enforced for using fire-pails for other purposes. During performances the fire-pump should be constantly kept under the steam pressure required to operate the same.

The safety appliances should be in charge of a special trusted inspector. He should, if practicable, have on the stage an office or watch-tower, from which, like the operator in a central-switch railroad tower, he can operate the fire-proof curtain, the stage-roof ventilators, the registers of the ventilators in the auditorium ceiling, the perforated-pipe system forming a water-curtain at the proscenium-opening, and the fire-alarm apparatus. He should also have telephone communication with the theatre-manager's office, with the engine or pump room, and with the nearest fire-department station.

The fire-proof treatment and impregnation of light dresses and gauze costumes should be insisted upon, particularly for the ballet-dancers, and the treatment should be renewed after each washing.

Besides the automatic alarms on the stage, in the manager's-room, and in the engine-room, a theatre should be connected by telephone and by fire telegraph with the nearest fire-station and with the headquarters of the fire-department.

The constant use of all safety appliances should be insisted upon. At every performance the fire-curtain should be lowered to insure its proper working in times of need.

Likewise should all theatre exits be thrown open and used nightly, so as to have the public become familiar with them. The oil-lamps in corridors and staircases should be lighted every night, and not extinguished until the entire audience has left the theatre. All exit doors should be plainly marked in large and distinctly legible letters. It is a good plan to mark all other doors, which do not lead to exits, either by the words "no exit," or else to designate them so that they may be easily recognized, as for instance, "toilet," "buffet," "cloak-room," "office," "ladies' retiring-room," etc.

The exits and staircases should be plainly shown on clearly printed plans of the theatre, hung up in conspicuous places in the foyers or corridors. The theatre plans and the exits should be printed on every theatre programme, and in a way so as to be clearly legible.

The number of persons admitted to a theatre should be strictly limited by law according to its seating capacity, and it is better still to license each division of the auditorium for a fixed number of persons.

No standing-room should be permitted, nor should camp-stools be used in the aisles. The open courts at the sides of the theatre should not be used for temporary storage of theatre trunks or stage scenery, but they, as well as all passages, should be kept clear from all encumbrances.

The utmost cleanliness should be maintained throughout a theatre, not only in the auditorium, but also on the stage, in the dressing-rooms, the toilet-rooms, and the places under the stage. The daily removal of all dirt, dry dust, sweepings, shavings, rubbish, oily rags, and other waste materials of all kinds must be performed with regularity. Pending removal, oily rags or cotton waste should be kept stored in metal boxes, closed by iron lids, and raised up from the floor on legs.

The theatre-director and stage-manager should at all times remember that the safety of the audience is the chief consideration. To accomplish this the strictest rules and regulations should be enforced, in order, first, to prevent the outbreak of a fire; second, to localize and confine a fire when it does break out; third, to protect the audience against fire and panic; fourth, to secure safe egress in case of fire or panic to the audience and to the theatre employés and actors; and fifth, to extinguish a fire in its incipiency, before it has a chance to spread and carry destruction.

The law should require every theatre-manager to have in the office of the theatre a complaint-book, which should be accessible to the public and to the press. In this book every person observing some real defect should call attention to the same, so that the complaint may be brought to the notice of the authorities for investigation and remedy.

No interior alterations affecting either the plan or the arrangement and construction should be made in any existing theatre without the approval of the building or fire departments.

Periodical Inspection.

A few words, in conclusion, about theatre inspections. Eternal vigilance forms the only assurance of continued safety in a theatre. There should be, to begin with, examinations and visits to all parts of the building after each evening and matinee performance. Then there should be general and frequent examinations, either by the architect who designed the structure, and who is more than any one else familiar with all its features, or by a specially appointed committee of experts. Such a committee would be suitably composed of the following members, viz., the manager, an architect, a builder, an underwriter or a member of the fire-department, an hydraulic engineer, and an expert electrical engineer.

A periodical inspection in detail of all gas and water fittings, frequent tests of the gas-pipes, of the fire-pump, of the automatic-sprinkler system, and of all other fire appliances; of the electric-lighting system, of the fire-telegraph and fire-alarm systems, and of the various other electric equipments; of the fire-proof curtain, the stage-roof ventilators, and the lightning-rod protection,—these are all necessary to prevent a failure in the working of any of the safeguards in an emergency.

There should be, moreover, occasional official inspections by the authorities, the fire or the building department, preferably without a previous announcement. Repeated inspections of new theatre buildings are necessary to prevent any transgressions of the theatre fire law by the theatre-manager, or the engineer in charge of the theatre, after the same has passed inspection and received the approval of the authorities. It also sometimes happens that after the theatre license has once been obtained internal changes in arrangement or equipment are made which would be contrary to the rules, and which would have a tendency to reduce the safety of the building.

III.

THE WATER-SERVICE AND FIRE PROTECTION OF THEATRES.*

EVERY effort is nowadays made to make theatre buildings as fire-resisting and safe as possible. Having been recently engaged in the preparation of plans and specifications, and in superintending the installation of the water and gas service plant and the fire-extinguishing appliances of several modern theatres, I have prepared a paper, giving a detailed description of the general features and fundamental requirements of the water and fire-protection service in theatres, in the hope that it may prove of use to architects, engineers, underwriters, and theatre-managers, and of general interest to all persons frequenting such places of amusement. By way of introduction it may be well to summarize briefly the chief elements which together constitute a safe and fire-resisting building.

General Requirements of Safety in Theatres.

The chief conditions of safety for theatre buildings are the following:

I. *As Regards the Location.*—The theatre to be, if possible, either isolated on all sides, or standing free

---

* This paper was originally prepared for the annual meeting of the New England Water-works Association, held in Boston, June, 1894.

on three sides. If located in a block, the building to have the full front on the public street, with a wide open court on either side. If placed on a corner lot, a wide court to be arranged at one side.

II. *As Regards the Plan.*—The chief divisions of the theatre to be entirely separate and distinct from each other. The stage to be separated from the auditorium, the dressing-rooms from the stage, the auditorium from the lobby, entrances, and staircases, by strong fire-walls.

The vertical subdivisions of the auditorium, the orchestra, dress-circle, balcony, and the galleries to be separated from each other by fire-proof floors and ceilings. Each division to have at least two separate and ample means of egress.

The stage to have at least two separate exits to the street.

The scene-docks, the carpenter's and painter's shops, and other workshops to be located in a separate fire-proof annex.

No living apartments to be planned in a theatre. No stores to be provided, except they are entirely isolated, fireproofed, and not in communication with the theatre.

III. *As Regards the Construction.*—The whole building to be constructed in a fire-resisting manner. All divisions to be enclosed by fire-walls carried above the roof. All staircases, corridors, floors, and ceilings and partitions to be built fire-proof. All roofs to be fire-proof.

The sides of the stage, the fly-galleries, the rigging-loft, and the stage roof to be built in a fire-proof manner. As little combustible material as possible to be used on the stage proper, and steel-wire ropes to be substituted for hemp ropes wherever practicable.

All dressing-rooms to be built fire-proof.

The boxes and fronts of balconies and galleries, the frame of the proscenium-opening, and the auditorium ceiling to be built of fire-proof material.

The main auditorium floor to be located on the street level, not in an upper story. All open stair wells to be avoided as dangerous, and all stairs to be enclosed in fire-walls.

IV. *As Regards the Interior Arrangement and Equipment.*—The opening in the proscenium-wall between the stage and the auditorium to be provided with a fire-proof iron or asbestos curtain. The apparatus for raising or lowering the fire-curtain to be on the stage, and not in the rigging-loft. Other openings in the proscenium-wall to be only at or below the stage level, to be as few in number as possible, and to be closed with automatic fire-proof iron or double tinned wooden doors.

The roof over stage to be provided with a large automatic stage ventilator or with sliding skylights, held closed by hemp rope run down to the stage level.

The central lustre in the auditorium ceiling and its ventilator to be abolished.

None of the windows in the building to be provided with iron grilles or gratings.

All seats to be securely fastened to the floor, and not more than thirteen seats to be placed between aisles. No camp-stools to be placed in the aisles.

All aisles, passages, corridors, and stairs to be kept free at all times from all obstructions. The aisles to have gradients, and not steps.

Separate stairs and exits to the street to be provided in sufficient number and width for each part of the audience, permitting the complete emptying of the theatre in from two to three minutes.

All stairs to have even and uniform treads and risers, and to be provided on both sides with strong handrails. No single steps or winding stairs to be permitted.

All doors to open outward and in such a manner as not to obstruct the passages.

Fire-escape balconies and outside iron stairs leading to the open courts or the street to be provided with at least two exits from each tier, and at least one such fire-escape to be located on each side of the auditorium. Iron fire-ladders for use of the firemen to be provided on the outside walls.

The theatre to be heated from one central heating apparatus. The steam-boilers to be placed outside of the building proper. The boiler smoke-flue to be lined with tile-pipe.

The lighting of the stage, and that of the passages, stairs, exits, lobbies, and rear portion of the auditorium, to be from separate gas-mains. Each of these two gas-mains to be controlled by shut-offs on the sidewalk.

The gas-meters to be located in well-ventilated fire-proof and isolated vaults.

All open gas-flames to be surrounded with large wire cages, and all bracket lights to be stiff. All footlights, border-lights, and bunch-lights to have special means of protection.

The lighting of foot and border lights to be done by means of electric flash-lighting, and not with alcohol torches.

Auxiliary oil-lamps or candles, protected against draft, to be provided in corridors, stairs, and exits.

All woodwork of stage, all decorative scenery, and all gauze costumes to be impregnated and rendered fire-proof.

The building to be protected against lightning by a well-equipped system of lightning-rods.

Finally, provision to be made for plenty of efficient fire-extinguishing appliances, always kept in readiness, and to include outside fire-hydrants, a fire-pump, lines of stand-pipes, with fire-valves in each tier of auditorium and of stage, and with fire-hose attached; an automatic-sprinkler system, with tank-supply and fire-department outside connection, casks of water, fire-pails, portable chemical extinguisher apparatus, pick-axes, fire-hooks, and other appliances, as described further on in detail.

V. *As Regards the Management.*—The fire-curtain to be used at each performance.

All doors, exits, and stairs to be kept open and used at each performance.

Doors not leading to exits to be marked with the name of the room to which they lead, or simply by the words " no exit."

All exits to be plainly marked in large legible letters.

After the performance lights in the auditorium not to be lowered until the audience has left.

The auxiliary oil-lamps or candles in stairs, corridors, and exits to be lighted before each performance, and not to be extinguished until the entire audience has dispersed.

The oil-lamps to be filled, cleaned, and trimmed by daylight only.

Safety lanterns in place of open lights to be used in entering large wardrobes.

Smoking in the theatre and in the dressing-rooms to be prohibited.

Strict rules to be enforced regarding careful use of fireworks and firearms on the stage. Paper wads for pistols and guns to be prohibited.

Open fires and open lights, candelabras with candles, and oil-lamps on the stage to be used carefully. The use of matches to be restricted to the kind known as " safety " matches.

The complete darkening of the auditorium during scene-shifting with raised curtain to be previously announced to the audience to prevent a false alarm or panic.

Renewal of chemical impregnation of all gauze dresses after each washing.

The fire-appliances to be kept in readiness, the fire-valves to be kept unencumbered, the fire-pails to be always kept filled with water, and access to the fire-valves not to be blocked.

The automatic fire-alarm to be kept in order.

The theatre to be in telegraphic and telephonic communication with the nearest fire-department station.

A fire-watch to be provided during performances, and also a permanent night-watch.

All oily rags and cotton waste to be stored in metal vessels with tight cover.

Frequent removal of all accumulation of oily waste, rubbish, paper and litter, dirt and dust.

Storage of large quantities of scenery on or behind the stage not to be tolerated.

Plans of the building to be hung up in conspicuous places in the theatre, and to be printed on all programmes, with the position of stairs and exits legibly marked.

Life-saving appliances to be kept on hand.

A fire-drill to be organized among the theatre employés and stage-hands.

No greater number of tickets of admission to be issued than that for which each division of the theatre is licensed.

No standing-room to be permitted in any aisles.

VI. *As regards Inspection.*—A nightly inspection of the building, including a visit to every part of the theatre after each performance, to be insisted upon.

The faithfulness of the night-watchman to be controlled by an electric watchman's clock.

The theatre to be frequently reinspected by its architect, by experts, by underwriters, or by the fire-department, the inspections to take place preferably at unexpected hours.

Periodical tests of all gas-piping, of all fire-apparatus, of the fire-alarm, telegraph connections, of the automatic stage-roof ventilator, and of the lightning-rods to be carried out.

Important as the observance of all these requirement is in the case of the newer and better constructed theatres, it is quite obvious that in the older theatres, which have been erected without any regard to fire-resisting qualities, they are, with the exception of those relating to fire-proof construction, still more indispensable to secure the safety of the audience and of the actors. As regards, in particular, the fire-service, which I am about to describe, it is self-evident that if it is needed in a theatre constructed in a fire-proof manner, it is absolutely indispensable as a protection for such structures as are flimsy in construction, highly combustible, and in many cases exceedingly hazardous.

Fire-extinguishing Appliances.

I will now take up the subject proper of my paper, and I propose to discuss in detail the fire-appliances needed for the extinguishment of fires in theatres. A theatre building may be planned and constructed in a fire-resisting manner and with all due precautions to prevent the rapid spreading of the flames, yet in spite of all conceivable measures of safety a fire may break out. Therefore, ample fire-appliances must be on hand and kept in constant readiness to fight the fire and to put out the flames. All fire-extinguishing apparatus should be of the very best of its kind, and it should always be properly cared for and kept in working order and ready for instant use.

A complete fire protection of a theatre must include:

1. Outside fire-hydrants;
2. Inside stand-pipes, with fire-valves, fire-hose, and fire-nozzles, supplied from direct street pressure or from—
3. Tanks; or from—
4. Fire-pumps;
5. An automatic-sprinkler system, with elevated water-tank and outside or fire-department connection;
6. A system of perforated pipes on the stage over the proscenium-opening;
7. Fire-pails and casks of water;
8. Chemical or portable fire-extinguishers;
9. A series of wet sponges on long poles;
10. Wet woollen blankets, asbestos sheets, and asbestos gloves;
11. Steam-nozzles or jets;
12. Fire-axes, fire-hooks, and knives.

Acknowledgment must in this place be made to the valuable services rendered to the cause of fire protection and fire extinction of all classes of risks by the researches and labors of the engineers of the Mutual Factory Assurance Companies of Boston, to many of Mr. C. J. H. Woodbury's papers on mill protection and on automatic sprinklers in particular, and to Mr. John R. Freeman's able specifications for fire-pumps, and fire-hose, and his elaborate and painstaking experiments on fire-streams and fire-nozzles. Taken together, these contain the most important facts upon the subject ever published, and the valuable conclusions reached by them may with advantage be applied to the equipment of fire-service plants in theatres.

1. Water-Supply and Water-Mains.

For the outside protection of theatres it is necessary that there should be a large water-main in the street or streets where the building is located. The size of the main should be at least 8 inches in diameter, while a 10 or 12 inch main would be still better. To this main should be connected the various outside hydrants for fire-department use, located in front of the theatre, in sufficient number and at suitable distances apart.

The outside fire-hydrants should preferably be frost-proof post hydrants, with four independent gate-valve connections, so as to give an opportunity, where the pressure is ample, to bring into play a number of hose streams. The hydrants should be examined frequently during freezing weather, and their working parts must be oiled occasionally with heavy mineral oil.

A very abundant supply of water for a theatre building is necessary for fire protection.

The size of the water-supply main run into the theatre is dependent upon the available pressure, and largely upon the number of fire-valves in the building where these are supplied under direct pressure, or else upon the size and capacity of the fire-pump installed. The New York Building Law, for instance, requires " the fire-pump to be of sufficient capacity to supply all lines of hose when operated simultaneously;" but in practice this requirement is never insisted upon, as it would lead, in even medium-sized theatres, to pumps, water-mains, and steam-boilers of unreasonable size. A " four-stream " fire-pump, delivering at the rate of 1000 gallons per minute, and able to supply from four to five fire-streams rated at from 200-250 gallons per minute each, would require a 12-inch suction-pipe. It would be next to impossible to secure from the city water-department a connection of this large size with the main. As a rule, a 4-inch supply main is all that is conceded, although a 6-inch supply or two 4-inch supplies from different street mains, and cross-connected within the building, are undoubtedly much to be desired. It is, therefore, evident that a reserve supply, stored in a large suction-tank, should never be omitted, and the smaller the service-pipe obtained, the larger should be the capacity of the suction reservoir.

The water available for fire protection must be ready under such a pressure as to permit the water to be forced in sufficient volume to all parts of the building. As a rule, in our large cities, the direct or street pressure is insufficient for fire-service. It therefore becomes necessary to supply the inside fire-streams either from sufficiently elevated open fire-tanks, or from closed tanks, operated under compressed air. But such open tanks, even if placed above the highest part of the stage roof, would fail to give a suitable fire-pressure (45 or 50 lbs.), except on the lower floors of the theatre. Closed tanks, operated by compressed air, require air-compressors and engines, and this means further complications in the mechanical plant of the building; therefore fire-pumps are usually substituted, and with an ample and sufficient steam-boiler capacity, and steam kept at proper high presssure, give in practice the best results as regards fire-pressure.

The four or six inch water-main is usually made of heavy cast-iron water-pipe, outside of the building, whereas in the building asphalted or galvanized wrought-iron pipe with screw-joints is used. It is necessary to place a shut-off valve on this branch outside of the building, so that in case of breakage of the inside water-main the water would not uselessly run to waste and flood the cellar, and also so as not to reduce the supply to the outside hydrants. Particular care should also be taken to run the inside main where it would be at all times protected against injury.

2. Water-meter and By-pass.

Theatres in this country are private enterprises, always erected, managed, and owned by individuals, firms, or corporations, and hence the water taken into the building is subject to the usual water-tax. In most cases the water-department insists on supplying the water by meter measurement. The water from the four or six inch main should therefore pass through a meter of corresponding size, and shut-off gate-valves must be placed on each side of the water-meter in order to be able to disconnect and exchange the same in case of repair. It is, however, in my judgment, not only unnecessary, but objectionable, to require that in case of an emergency the fire-pump should draw its supply through a meter. In several theatres, the fire protection of which was under my charge, I therefore endeavored to obtain from the city water-department a full-size by-pass, controlled by a gate-valve. My intention was that this by-pass valve should ordinarily be kept strapped and sealed with a seal of the water-department to prevent the unlawful use of the valve, and that it should be cut and the valve opened only in case of fire, so as to admit a direct supply to the fire-pump. I learn that such by-passes are granted for this purpose in France and Germany, but I was unable to convince the New York City Water Department of the desirability of such a device, and the by-pass arrangement proposed did not receive the approval of the authorities, although no good reason was given for the refusal.

3. SHUT-OFFS AND GATE-VALVES.

All supply-mains and the distributing service-pipes in a theatre should be controlled by valves made of gun-metal, in the best manner, and only the best quality of valves should be purchased. Globe-valves should not be used, gate-valves being much preferable, because they have a full waterway. At the pumps angle-valves may be used on the suction-pipes.

The fire lines starting at the fire-pump and connecting with the stand-pipes should preferably not have any shut-off valves, or else if valves are used they must be kept open and strapped; otherwise it may happen in a case of emergency that on turning open a fire-valve no water will be had, the gate-valve on the line having been accidentally closed in the cellar. All valves should be provided with wheel handles, and these should be marked with arrows indicating the direction in which they open. It is also recommended to provide all water-valves with telltales showing automatically at all times and at a glance whether the valves are open or closed. Such telltale indicators may be obtained from many of the firms making the manufacture of gate-valves a specialty, and among the best are those of the Chapman, Ludlow, Kennedy, and Walworth Manufacturing Companies.

4. FIRE-PUMPS.

The fire-pumps installed in theatres are either rotary steam-pumps of the fire-engine type, or else direct-acting steam fire-pumps. Excellent rotary steam fire-pumps are made by manufacturers of steam fire-engines, but in theatres the direct-acting pump is used to a much greater extent. While the ordinary so-called fire-pumps will answer the purpose, it is advisable to use in theatres the best procurable kind of pump. Of these the "Underwriter pattern" steam fire-pump is without doubt the most advanced and most serviceable type of direct-acting pump. I will give a brief statement of its chief features, based upon the admirable and elaborate specifications prepared by Mr. John R. Freeman, Hydraulic Engineer to the Associated Factory Mutual Insurance Companies, to which I refer for further details.

The "Underwriter" fire-pump is a pump of the "Duplex" type, built in a very heavy and substantial manner, and fitted with certain improvements. This pump differs from the ordinary direct-acting fire-pumps in the following points, viz.:

1. The steam-ports and the water-passages are made more direct, and from 25 to 50 per cent larger; the suction-pipe and the air-chambers are increased in size correspondingly, so as to make the pump more powerful and able to deliver a larger volume of water with-

out water-hammer, and so as to permit the pump to be run at high speed.

2. The principal working parts of the pump are made rust-proof, so as to permit the pump to start instantly after long periods of disuse. Thus the piston-rods and valve-rods are made of Tobin bronze instead of steel, and the water-plungers and stuffing-boxes are made of brass instead of cast iron.

3. The shells, castings, and the bolting are warranted especially strong, to withstand heavy pressures, and the whole pump is accordingly built heavier.

4. The pump is provided with numerous special attachments, such as a large vacuum chamber, a water-pressure gauge, a steam-pressure gauge, a water safety or relief valve, a set of brass priming-pipes and special valves, a number of hose-valves on delivery end of pump, a stroke gauge, a sight-feed lubricator, and a capacity plate, all of which fittings are not included in the ordinary style of fire-pumps.

Fire-pumps are usually required to start at a moment's notice and to run at double the piston speed of boiler-feed or house pumps, and the above requirements are found in practice necessary to enable the pumps to run without jar or water-hammer.

Underwriter fire-pumps are now manufactured by all the principal manufacturers of direct-acting pumps, and nearly all of them have adopted a 12-inch stroke as preferable, although a few still adhere to a 10-inch stroke.

There are five sizes of Underwriter pumps made, and the chief dimensions of these are given in the following table:

TABLE I.

SIZES OF UNDERWRITER FIRE-PUMPS.

Designation of Pump.	Capacity in Gallons per Minute at Full Speed.	No. of Hose Valves on Pump.	Diameter of Steam Cylinder, Diameter of Water-Plunger, Length of Stroke.	Ratio of Piston Areas.	Full Speed for Rating Capacity.	Diameter of Suction-pipe.	Diameter of Discharge-pipe.	Diameter of Steam-pipe.	Diam. of Exhaust-pipe.	Size of Safety- or Relief-valve.	Capacity of Vacuum-chamber.	Capacity of Air-chamber.	Boiler Horse-power and Steam-pressure required.	Use of Pump.
One-stream pump...	250–320	1	12 × 6 × 12 inches	4 to 1	For 10 in. st. 75 rev. per min. 125 ft. pis.tr	6 in.	5 in.	2½ in.	3 in.	2½ in.	8 gal.	10 gal.	50 H.P.	This is too small for a fire-pump. Is a good size for a boiler-feed and domestic supply.
Two-stream pump..	500	2	14 × 7 × 12 inches or (16 × 8 × 10 inches)	3 to 1	160 "	8 in.	6 in.	3 in.	4 in.	3 in.	13 gal.	17 gal.	80–100 H.P. 40 lbs steam.	This is the ordinary size for small mills.
Three-stream pump	750	3	16 × 9 × 12 inches, or (16 × 9¼ × 10 in.)	3 to 1	140 "	10 in.	7 in.	3½ in.	4 in.	3½ in.	18 gal.	25 gal.	115 H.P. 45 lbs. steam.	This is the ordinary size for general use.
Four-stream pump.	1000	4	18 × 10 × 12 inches, or (18⅞ × 10⅞ × 10 in.)		64 "	12 in.	8 in.	4 in.	5 in.	4 in.	24 gal.	30 gal.	150 H.P. 44 lbs. st.	This is the size for large factories.
Six-stream pump...	1500	6	20 × 12 × 15 inches, or (20 × 12 × 16 in.)	2¼ to 1	15 "	14 in.	10 in.	5 in.	6 in.	5 in.	30 gal.	40 gal.	200 H.P. 50 lbs. steam.	This size is sometimes used for large factories, but two smaller pumps are generally better.

Details of Underwriter Fire-pump.—In all except the largest size of fire-pumps the ratio of steam-cylinder area to water-plunger area is as 3 to 1, but where the available steam-pressure is low the ratio should be as 4 to 1.

In the above table 250 gallons is taken as the standard allowance for a good $1\frac{1}{8}$-inch smooth-nozzle fire-stream, and this, for a fire-pump supplying a sprinkler system, is equivalent to from 15 to 20 automatic sprinklers.

All fire-pumps should be built of sufficient strength to stand a steam-pressure of 80 lbs. with steam-valves wide open, and giving, with all water-valves closed, a water-pressure of 240 to 320 lbs. per square inch. The pump must be well and strongly bolted in all its parts, and it must be so designed and the strength of the principal dimensions so calculated that a full head of steam may be safely turned on to a cold pump without cracking or breaking same by unequal expansion.

The piston and valve rods should be of solid Tobin bronze, and the water-plungers must be of solid brass or bronze. All valve-seats to be of U. S. gun-metal composition in order to avoid the sticking of valves. The suction-valves should have a combined area equal to 56 per cent of the plunger area, for pumps running at full speed and with 12-inch stroke. For a 10-inch stroke this proportion should be 50 per cent and for a 15-inch stroke 64 per cent.

The force or discharge valves should have two thirds of the total port area.

The steam-port should have not less than $2\frac{1}{2}$ per cent the area of its piston, and the exhaust-port not less than 4 per cent. Cushion-valves should be provided to regulate the amount of steam-cushion, and a stroke-gauge to indicate the length of the stroke.

When the pump draws from an open reservoir only, not over 20 feet away, through a large suction-pipe, the suction or vacuum chamber may be omitted. The air-chamber on the discharge end is quite necessary in the case of fire-pumps, so as to avoid the constant vibration of the fire-hose, which causes the hose to wear out quickly. It is usually built of cast iron, and should be tested under 400 lbs. hydraulic pressure. It is well to paint this air-chamber inside and outside to reduce its porosity. An air-chamber of hammered copper is preferable on many accounts, although more expensive, and should be tested under 300 lbs. pressure.

Each fire-pump must have a capacity plate, usually attached to the air-chamber, and made of enamelled iron, with large black or blue letters on white ground.

There should be on discharge end of pump, close to the air-chamber, a water-pressure gauge, with one-fourth-inch lever handle-cock, and attached to the steam-chest inside of the throttle-valve a similar steam-pressure gauge.

All fire-pumps should be fitted up with a safety relief valve of Ashton or Crosby's make and pattern, which is attached to delivery end of pump, preferably extending horizontally in board of air-chamber. The hand-wheel for regulating the pressure should be within easy reach, and should be conspicuously marked with arrow indicating direction of opening. The safety-valve is ordinarily set at a working pressure of 100 lbs. per square inch, and its size should be such that if pump runs at 125 lbs. pressure and two-thirds speed, the valve discharges all the water freely. This discharge should be by a vertical downward pipe one foot long, into a funnel with 6, 8, or 10-inch waste-pipe, according to size of pump. The discharge-pipe is open at the funnel in order to indicate to the engineer running the pump the water wasting through the relief-valve.

Each fire-pump must have all necessary $\frac{3}{8}$-inch bore brass drip or drain cocks, all provided with lever handle. Of these nine are required, as follows: One at each end of the two water-cylinders, one at each end of the two steam-cylinders, and one above the upper deck-valve.

Four $\frac{3}{8}$-inch brass lever handle air-cocks are to be attached to the pump, one at top of each of the four plunger-chambers for use in priming the pump. These air-cocks should preferably be fitted with check-valve which permits outflow and prevents influx of air when the plunger is sucking.

Each fire-pump should be fitted with a set of 1-inch brass priming-pipes, including one 2-inch main valve and four suitable check-valves leading one into each of the four plunger-chambers. The priming-pipes must lie close to the pump so as to be out of the way of the hose.

It is preferable to feed these priming-valves from a 100 to 200 gallon feed-tank, so that if the city water is shut off the pump is not disabled.

Where an efficient set of priming-pipes is provided,

a foot-valve on the suction-pipe of the pump is not needed.

A number of straightway brass fire-valves with hose screw at end suitable for standard 2½-inch fire-hose coupling, are provided at delivery end of each fire-pump.

All fire-pumps should be tested before acceptance, by a number of tests, for delivery, capacity, strength, etc.

A fire-pump, set up in good order, properly packed, must run smoothly, noiselessly, without jarring, slamming, jumping, hammering, at its full rate of speed (70 revolutions per minute) with full length of stroke, and maintain a water-pressure of 100 lbs. with from 45 to 50 lbs. steam-pressure. The water during the tests should be discharged through lines of 2½-inch cotton, rubber-lined fire-hose, each line about 150 feet long, and through 1⅛-inch standard Underwriter smooth fire-nozzles, the hose being connected to the valves at the pump.

A two-stream pump should be tested with two lines of hose, a four-stream pump with four lines, and so forth. The quietness of the hose is an indication of the perfection of the pump, and bad pulsations indicate a non-uniformity of delivery.

The test for strength is made by shutting nearly, but not quite, all water outlets, so that the pump moves very slowly, and screwing the safety-valve down hard. If steam is admitted sufficient to give 240 lbs. water-pressure, all pump joints should remain tight.

Fire-pump for Theatres.—As regard the size of fire-pump required for a theatre, the minimum size for use in small theatres would be a two-stream pump, having a capacity of 500 gallons per minute; for medium-sized theatres I would recommend the use of the four-stream pump, supplying 1000 gallons per minute, and for large theatres either the six-stream pump (capacity 1500 gallons), or, better, two four-stream pumps (combined capacity 2000 gallons).

Although a large theatre may be provided with from 25 to 50 fire-valves, the requirement that the pump should supply all the streams simultaneously, as it reads for instance in the New York Building Law, seems to me altogether unnecessary, and in practice it is never insisted on. As is well known, most of the theatre fires have their origin on or about the stage, and all that can be reasonably required is that the half a dozen or so fire-valves operated at these points should be supplied with a full and effective stream. If the fire has already gained so much headway that it cannot be checked or controlled by the hose-streams on the stage, persons would rarely be found who would risk their lives by staying in the burning building to handle additional streams. At such times the building had better be turned over to the city fire-brigade and their fire-engines. The chief object of the inside fire-valves is to fight local fires, either on the stage, in the flies, in the dressing-rooms, in the boiler-room, or in some part of the auditorium, before they have gained any considerable headway.

The fire-pump of a theatre must under all circumstances be kept ready for instant operation during all performances. Sufficient steam-pressure must be maintained in the boilers for this purpose. The fire-pump must be in good order at all times, and should be used at least once a week.

The fire-pump in theatres is, as a rule, used to supply the fire-tank of the sprinkler system, and a $2\frac{1}{2}$ or 3 inch pump-riser, which has a shut-off valve at the pump, is carried up to this tank. Since the tank of the sprinkler system does not require refilling so often, it is well to make it a practice to use the fire-pump occasionally for filling the house-tank by cross-connecting its discharge end with the pump main for the house-pump.

The discharge outlet of the pump is fitted up with a manifold, to which all the $2\frac{1}{2}$-inch fire stand-pipes in the theatre are separately connected, without any gate-valves at the manifold.

An excellent arrangement, which for instance in New York City is made a requirement of the fire and building departments, is that the fire-pump is fitted with an automatic pressure-regulator, so that in case a fire-valve is opened anywhere the fire-pump is started instantly and automatically. I have used for this purpose with success the Fisher pressure-regulator or pump-governor.

The fire-pump should be placed either in the engine-room or in the boiler-room, where in the event of a fire in the theatre it would be least liable to be affected by the flames. It should be set on proper brick foundations, finished on top with an axed bluestone cap, set perfectly level. The smaller sizes of pumps should be bolted down at the water end only, leaving the steam end free for expansion; but in the case of the larger sizes of fire-pumps all bolting and anchoring to the foundations is usually omitted, for the large weight of the pumps is sufficient to keep them securely in place.

The steam-boiler should be of sufficient capacity to furnish ample steam under high pressure to the fire-pump. The boiler should be fed from injectors and from a good-sized boiler feed-pump, to avoid the risk of a boiler explosion in case the water in the boiler should get low when the fire-pump is going at full speed.

5. SUCTION-RESERVOIR.

Ordinarily, the fire-pump, as well as the boiler and house-tank pumps, should draw their supply from an open suction-reservoir, and not directly from the supply-mains. This suction-tank, if suitably large, would act in case of a fire as a storage-reservoir for the fire-pump. It is therefore advisable to make this tank as large as the space at disposal and the available means permit. A 2000-gallon tank should be considered the minimum size, but where it can be obtained, I should advise using at least a 10,000-gallon tank. This would give to a four-stream fire-pump a reserve supply lasting at least 10 minutes, with the pump going at full speed.

The suction-reservoir is usually made of boiler-iron sheets, well and tightly riveted together, and strongly stayed and braced. It is supplied from the theatre water-main by a number of ball-cocks of large size. To guard against overflow in case the ball-cocks get out of order it is provided with a large overflow pipe, carried over a trapped and sewer-connected sink. The suction-tank is connected with the fire-pump by a suction-pipe of diameter corresponding to the size of the pump, and it is also connected with the other steam-pumps constituting together the pumping-plant.

The suction end of the fire-pump is also connected with the direct supply fom the street main. It is advisable to cross-connect all pumps in this manner, so that in case the reservoir is emptied, for cleaning or other purposes, the pumps are enabled to draw water directly from the street. The suction-pipe of the fire-pump varies from 8 to 12 inches, according to size; and where it connects with the street main, which as stated above is rarely larger than 4 inches, a suitable reducing fitting should be used, and the use of bushings should not be permitted.

A still better fire protection would be obtained by providing two independent supplies to the fire-pump, from mains on different streets, so that in case the fire-engines have arrived and being connected to the fire-hydrants outside of the theatre exhaust the capacity of the main in front of the theatre, the inside fire-pump would still be kept supplied with water by the second independent main.

6. Fire Stand-pipes.

At suitable points in the four subdivisions of a theatre, fire stand-pipes, with fire-valves and hose attached, should be placed. The largest number of fire-valves is required on or about the various levels of the stage where a fire is most likely to break out. There should also be fire stand-pipes in the auditorium, in the corridors and lobbies, and near the dressing-rooms, and they should be so distributed that the remotest corner can be reached by the fire-hose at each of the outlets.

The number of stand-pipes required and their position depend, therefore, upon the size and area of the theatre and upon the length of the fire-hose attached to the fire-valves. The New York City Building Law requires that one stand-pipe be provided on each side of the stage, with outlets on every floor and gallery, from the under-stage to the rigging-loft; one stand-pipe on each side of the auditorium, with outlets in each tier from the cellar to the gallery; one in the property-room, and one in the carpenter's shop. Thus, in the new Harrigan Theatre there are four lines, and in the new Fifth Avenue Theatre five lines, of stand-pipes. Larger theatres would require a still greater number.

The stand-pipes vary in size from 2 to 3 inches, but it is advisable to make $2\frac{1}{2}$ inches the minimum size. All standpipes should stand entirely free from the walls. Each line should be connected directly with the fire-pump in the engine or boiler-room, and, as already stated, there should be no shut-off valve at the pump. The usual material for stand-pipes is galvanized wrought iron. In view of the fact that the pipes have to stand an internal fire-pressure of about 100 lbs. per square inch and sometimes strong shocks and jars from water-hammer, it is advisable to use double or extra heavy wrought-iron pipes. Tinned and annealed brass pipes would, undoubtedly, be preferable for fire stand-pipes on account of their greater interior smoothness, but the cost of brass pipe generally percludes their use.

The vertical stand-pipes should run as straight as possible, avoiding offsets which mean increased friction. Likewise should the horizontal lines connecting the standpipes with the fire-pump be run as direct and with as few changes in direction as possible. Where changes are necessary, it is far better to use 45° elbows, and right-angle elbows should be entirely avoided. The New York Building Law stipulates that the fire stand-pipes must be connected only with the fire-pump, and not with the fire-tank which supplies the sprinkler-system.

In many theatres of Europe provision is made for large storage-tanks placed over the highest part of the stage, and filled partly with water and partly with compressed air of at least 75 lbs. per square inch pressure, and in this case the fire-valves and stand-pipes are supplied from these special fire-tanks, which also may supply a system of perforated pipes placed over the stage, which in case of an emergency serve to deluge the stage and stage scenery with water.

7. Fire-valves.

At each outlet of the fire lines, and also at the fire-pump, there should be provided a fire-valve. Straightway gate-valves of best gun-metal, such as made by the Chapman Valve Manufacturing Co., the Ludlow Valve Co., and others, are the best, and they are preferable to angle globe-valves, although the latter have the advantage of permitting the hose to hang down straight, and thereby avoid the possible "kinking" of the hose at the valve outlet. Fire-valves should be of the best obtainable make, as those of poor quality and bad workmanship are liable to become stuck in the seats.

The size of the fire-valve varies from $1\frac{1}{2}$ to $2\frac{1}{2}$ inches, and in cities where this matter is regulated by building laws the standard fire-department size, viz., $2\frac{1}{2}$ inches, is called for.

The number of fire-valves varies according to the number of stand-pipes, the number of tiers or galleries in the audience, and the number of fly-galleries over the stage. Thus, at the Fifth Avenue Theatre in New York City, 25 fire-valves have been provided, located as follows, viz.: two in parquet, two in balcony, two in gallery, one on each side in each case, making in all six for the auditorium; two under the stage where the traps are, two on the stage level, one on each side of the proscenium opening, four in the two fly-galleries on each side of stage, and two in the rigging-loft, making ten in all for the protection of the stage; and one in basement smoking-room, one in engine-room, one in cellar near supers' dressing-rooms, five in the corridors adjoining the actors' dressing-rooms, on the various tiers, and one in the costumer's room on the top floor.

Large-sized theatres and opera-houses would require a still greater number.

All fire stand-pipes must be kept free from all encumbrances and readily accessible, and nothing should be hung on the fire-valves or the hose-racks. Where these lines are not bronzed with silver, gold, or aluminum bronze, it is desirable to paint the fire stand-pipes a bright-red color to make them conspicuous. A good plan consists in attaching indicating pressure-gauges at some or all outlets of the fire stand-pipes, which show if the available pressure is sufficient for fire purposes.

8. Fire-hose.

Only the best kind of hose is suitable for use in a theatre, and no expense should be spared in this direction. It is true that often the cheapest kind of fire-hose, with rough inside and only made "to sell," is provided, merely to comply with the requirements of the law or of the underwriters. But a poor quality of hose, hung up for inspection or exhibition purposes only, will invariably prove worthless, and leak like a sieve or burst when needed in an emergency.

The fire-hose may be leather hose, or solid rubber hose, or linen or cotton hose lined with rubber, or, finally, unlined linen hose.

Leather hose is very expensive, besides being stiff and unwieldy, and for these reasons it is not much used, and the same may be said of solid rubber hose. The choice for theatres lies between unlined linen and rubber-lined cotton hose.

The New York Building Law does not state the kind of hose required, but in the actual inspection of new theatres the department requires all theatre fire-hose to be rubber-lined three-ply fire-department cotton hose, $2\frac{1}{2}$ inches inside diameter, able to stand 300 lbs. pressure per square inch, and it rejects unlined linen hose, even of the best brand.

Personally, I do not concur in this view. Good woven linen hose can be obtained which is practically water-tight except for a slight sweating under pressure at the beginning, and which is able to stand a bursting strain of from 300 to 400 lbs. I should favor, for indoor use in theatres, the use of unlined linen hose, which is not so heavy and difficult to handle as rubber-lined hose. Carefully woven hose may be obtained in brands which have as much smoothness of the inside as is ordinarily required. While rubber-lined hose may be tighter, and therefore will prevent unnecessary damage by leakage of water when new, it is my judgment that when hung up in a theatre, and perhaps not used for a long period, the rubber lining will crack or disintegrate, and in such condition the hose will be less useful than a good brand of carefully woven unlined linen hose, which if kept dry will last for many years.

In this view I believe I am in accord with all experts who have made a special study of the question of suitable material for fire-hose, the universal verdict being that "however suitable rubber-lined linen, rubber, and leather hose may be for public fire-departments, both experience of use and tests show that they are not so well adapted for inside use." The chief argument in favor of rubber-lined hose, namely, that it is not cut or injured when lying over sharp stones, and that it is not worn by horses' hoofs or wagon wheels, does not hold good in the case of fire-hose as applied to the inside fire-valves in theatres.

Moreover, while I am fully cognizant of the great loss by friction occurring in fire-hose smaller than $2\frac{1}{2}$

inches in diameter (the loss by friction in a 2½-inch hose is stated by Mr. Freeman to be only one third of what it is in a 2-inch hose), I am inclined to favor for inside use in theatres, as well as in other public buildings and institutions, the use of 2-inch, or better even of 1½-inch, hose, together with a smaller size of fire-nozzle. Such an outfit would be lighter, more compact, and more convenient for quick use. All inside fire-hose in theatres is primarily and chiefly intended for use in an emergency by the theatre employés, who are seldom, if ever, trained firemen. It is intended for putting out quickly a small fire during its first stages. When a fire in a theatre has once reached a stage where it calls for the action of the city fire-department, the heat and smoke of a fierce fire would be so great as to drive the men back, and render the inside fire-valves nearly or quite useless. The standard 2½-inch rubber-lined or even the unlined cotton hose are much too heavy to be handled with success by ordinary employés. Few men, indeed, realize that it requires the full strength of two firemen to hold a standard fire-nozzle under a water-pressure of only fifty pounds, and that an inexperienced man cannot hold a fire-nozzle if the pressure exceeds 25 or 30 lbs.

At each fire-valve there should be provided at least fifty feet of fire-hose, and the same must be kept always attached to the hose end of the fire-valve. It is important that every corner may be reached by a line of fire-hose, and the location and number of fire standpipes and fire-valves should be determined with this object in view.

9. Fire-nozzles and Couplings.

The fire-hose should be provided with well-fastened couplings. If the hose is 2½ inches in diameter, the couplings should be of standard fire-department size, shape and thread, and the couplings must not be obstructed by washers projecting on the inside.

Each length of hose must be provided with a well-finished strong brass nozzle. There are numerous shapes and patterns of nozzles, or "play-pipes." Whatever pattern may be chosen, the inside of the nozzle must be absolutely smooth, and as true as a gun-barrel. Smooth, plain nozzles are the best, and there is no advantage in using ring nozzles. Mr. Freeman's experiments on fire-jets and fire-nozzles established the fact that the discharge stream from a ring nozzle is 25 per cent less than that from a smooth, plain nozzle.

Short nozzles, about 1 or 1½ feet long, are not so easily handled as long nozzles, and their shape is not such as to throw the most efficient jet. It is always better to provide swivel handles at the base of the play-pipe, these being useful not only in holding the nozzle, but also in carrying the hose up a ladder or in shifting hose. The best form of nozzle for 2½-inch hose is probably that designed by Mr. Freeman, and which may be obtained from all dealers in firemen's supplies, under the name of the "Underwriter," or "Factory Mutual," pattern of play-pipe. The principal dimensions of such play-pipes are: Length over all, 30 inches; diameter at base of play-pipe, 2½ inches; diameter at base of standard nozzle, 1¾ inches; diameter of opening of nozzle, 1⅛ inches.

It is best to make the play-pipes and nozzles of polished brass or gun-metal, and it is a good plan to have the body of the play-pipe wound outside with cord to secure a better grip and to prevent coldness to the touch. Rubber and cotton fabric play-pipes are not as durable, for both the rubber and the cotton are liable to rot.

When a $1\frac{1}{2}$-inch hose is used for inside protection the fire-nozzle opening need not be larger than $\frac{7}{8}$ inch.

Beware of cheap nozzles, made of light or thin brass, rough on the inside, and which easily become dented, battered, or bruised.

10. MONITOR NOZZLES.

A very excellent device for fire protection, and one which seems to be particularly adapted for the fire protection of the stage, the fly-galleries, and the rigging-loft, is the so-called "Monitor Nozzle," invented by Andrew J. Morse of Boston. It consists substantially of a play-pipe attached to a revolving chamber, and provided with means for lowering or raising the play-pipe, thus permitting the hose-pipe to be inclined or elevated at any angle desired, and to be turned into any direction horizontally. Such device would enable the attendant to cover any desired point with the fire-stream.

This nozzle is similar in construction and operation to the large and very powerful nozzles now in use on the recently built fire-boats of the New York Fire Department.*

---

* Since writing the above I learned from Messrs. Morse & Son that the powerful "Monitor nozzle," throwing a solid $5\frac{1}{4}$-inch stream, with

One feature of great importance is that if the Monitor nozzle is once pointed in any particular direction, it is so balanced that it will remain in this position without any attention whatever. This would enable a single man to get an effective stream on a fire, and while he leaves it under operation, to go to another nozzle and thus start a second or several more streams, or else to give the fire-alarm, or to attend to other duties in connection with the saving of life or property during a theatre fire. Moreover, the necessary delay incident to the getting out of the fire-hose is avoided, and thus valuable time at the beginning of a fire is saved. This, as experience teaches, often means the saving of the building, from total destruction. Finally, although these Monitor nozzles are expensive, they render the use of fire-hose, which at best lasts not many years, unnecessary, or at least they reduce the amount of fire-hose needed in a theatre.

11. Hose-Racks.

The fire-hose attached to each fire-valve must be hung up or supported in such a manner that it can be readily run out without any twists or kinks, and quickly stretched out on the floor ready for use. There are several devices in the market which accomplish this.

In the swinging hose-reels the hose is reeled around, while in the Guibert swinging hose-rack the hose is

---

which the fireboat "New Yorker" has recently been fitted up, has been manufactured by them, and is of the same construction as the $2\frac{1}{2}$-inch "Monitor nozzle" recommended by me for fire protection of theatre stages.

folded in layers. Either device may be attached to a wall-plate, or else by means of pipe clamps directly to the stand-pipe, where this stands free and a few inches out from the wall. In the racks rubber-lined hose is more difficult to put than unlined hose, and there is danger of the hose cracking in the folds, but for unfolding quickly a length of hose, the rack is undoubtedly superior to the reel. Sometimes the hose is simply hung up on brass hooks or pegs, but this is a crude device, and involves the loss of much valuable time in stretching out the hose.

A better device is the saddle rack, designed by the Boston Woven Hose and Rubber Co., which takes up less space than other racks and permits the hose to come off easily and without catching.

Other special devices are Schenck's swinging hose-reel and the Peerless stationary and swinging reels. They are used extensively on the Pacific coast, but the Schenck hose-reel may be obtained in Philadelphia, Boston, or New York from Eastern agents.

The swinging hose-racks, so largely employed for unlined linen hose, are made of iron and usually japanned a bright red; but where desired, in conspicuous locations, they may be made of solid brass, either polished or nickel-plated. If intended for rubber-lined cotton hose, the racks must be made extra strong and extra deep.

12. HOSE-SPANNERS.

At each fire-valve there should be provided at least one hose-spanner. Of these there are several good patterns in the market, and among the best are the

Allen and the Taber patent spanners which will turn a coupling to the right or left without removing the spanner from the coupling.

13. AUTOMATIC-SPRINKLER SYSTEM.

We have in the foregoing considered one essential form of fire apparatus for the protection of theatres, namely, the fire-pump, with its direct street supply and suction connection from a reserve reservoir, with its connections to numerous fire stand-pipes scattered over the theatre, and each provided with a number of fire-valves with fire-couplings, fire-hose, play-pipes, hose-racks, and spanners, or with Monitor fire-nozzles. The theatre employés should be entirely familiar with this apparatus, and must be thoroughly drilled in its use.

We must next consider a second essential system of fire protection of the stage, consisting in the equipment of an automatic-sprinkler system.

The automatic sprinklers which are now so universally fitted up in mill buildings, factories of all kinds, and in warehouses have also been introduced some years ago in the theatres of the United States, and there are instances on record where the timely opening of such sprinklers at the outbreak of a theatre fire saved the building from destruction.

On January 6, 1891, fire broke out at the Opera-House at Woonsocket, R. I., the cause being a front border light igniting a gauze drop, from where the flames jumped to the drop-curtain. The stage was protected with Grinnell's automatic sprinklers, under a pressure of 100 lbs. from the city water-supply. Under the gridiron and just above the drop-curtain six sprinkler-heads opened, also three more under the roof, and five more under the stage ventilator. The fire was quenched completely by fourteen sprinklers.

At the fire at the Union Square Theatre on Feb. 28, 1888, the Gray sprinklers saved the building.

At the New Theatre Royal at Bolton, England, a gas explosion was the cause of a fire which was put out by automatic sprinklers.

In November, 1882, a fire was extinguished in the Providence Théâtre Comique during a play, by automatic sprinklers, with so little injury, even to the scenery, that the performance was not stopped.

I do not know of any European theatre fitted up with automatic sprinklers, except the Theatre Royal at Bolton, just mentioned. The perforated-pipe system and a water-curtain, of which I shall speak further on, take there the place of the automatic sprinklers.

A sprinkler system is usually installed in only a portion of the theatre, viz., in that portion of the building lying back of the proscenium-wall, where the largest amount of highly combustible material is found, and designated as follows: Under the stage roof, under the gridiron or rigging-loft, under all fly-galleries, and under the stage; to which is often added the scene-docks, the property-room, the carpenter and painter's shop, all actors' dressing-rooms, the rooms for the supers and the ballets, and portions of the basement.

The loft over the auditorium is seldom piped for sprinklers when the space is not used, nor do the rules of the Board of Fire Underwriters require this.

But where this loft is used for storage, particularly of combustible material, it is better to install sprinklers.

As is well known, there are three principal sprinkler-pipe systems, namely, (1) the ordinary wet-pipe system; (2) the non-freezing wet-pipe system; and (3) the dry-pipe system.

The ordinary wet-pipe system is the one in most general use. It is open only to two objections, first, that in buildings not constantly or sufficiently heated it is liable to freeze, and thus become inoperative just at a moment when wanted, or to do damage by water by the bursting of the frozen pipes; second, that in the event of a sprinkler head becoming defective or damaged and opening it is liable to cause damage to the building or its contents by flooding same with water. It may be stated that the best sprinkler heads in the market are now so perfected as to almost preclude such an accidental bursting of a head.

The second or non-freezing wet-pipe system has an open water-tank and a small air-tight iron tank feeding the pipes, which are filled with a non-freezing and non-corrosive liquid. It is used where the building in which the sprinkler system is installed cannot be heated. The two tanks are connected by a siphon, so that when a sprinkler opens the system is fed from the water-tank. This system is open to the objection of possible leakage, and, moreover, a possible corrosion of the sprinkler heads.

The third or dry-pipe system is extensively advocated for buildings in which water-pipes would be liable to freezing. There are several modifications of this system, the pipes in one system being filled with compressed air instead of water. In all of these there is great difficulty in making the system absolutely airtight. Air-pumps are generally provided to keep the air-pressure up, and the system must be frequently watched. An improved dry-pipe system has, in addition to the sprinkler pipes and heads, an independent air-pipe, with fusible solder-plugs, and when one of these is operated by the heat of a fire an automatic valve is opened admitting the water from the tank to the pipes.

In theatres the temperature can be at all times kept up to a sufficient degree of heat to insure the non-freezing of the water in the pipe system, and therefore it is the rule to install the ordinary wet-pipe system.

It is the purpose of an automatic-sprinkler system to extinguish a fire in its incipiency, through the agency of the fire itself. A fire if started opens up only a limited number of sprinkler heads; hence there is no useless waste of water, no unnecessary damage, and the water discharged is concentrated in the spot just where wanted.

A wet-pipe sprinkler system comprises a wooden Cooper's tank, placed on the highest part of the stage roof, and of a capacity corresponding to the number of sprinkler-heads. The installation for a theatre requires generally from 100 to 300 sprinkler-heads, suitably distributed; several $1\frac{1}{2}$, 2, and 3 inch risers; a separate shut-off valve for each floor or tier; a $2\frac{1}{2}$ to 3 inch riser from the pump to the tank; a watchman's automatic fire-alarm with large gongs—one on the stage, the other on the outside of the building; a low-water alarm with indicator in the pump or engine-room; one or more 3-inch pipes running from the main riser to the outside of the building, to form a fire-department connection (for auxiliary supply), which is provided with check-valve.

It is essential that all sprinkler systems should have two independent approved sources of supply, of which one should be always turned on. Speaking generally, the supply to the sprinkler system may be either from a roof tank or from reservoir pressure, or from direct pressure from public street mains, or from special sprinkler fire-pumps, or from fire-department steam-engines by means of the outside fire-department connections.

The following resulting combinations for a double supply are usual, viz.:

1. Primary supply: The street main.
   Secondary " An elevated fire-tank on the roof.
2. Primary supply: The street main.
   Secondary " An automatic steam fire-pump with steam always kept on.
3. Primary supply: An automatic fire-pump.
   Secondary " A roof tank.
4. Primary supply: Roof tank.
   Secondary " Rotary pump.
5. Primary supply: A roof tank.
   Secondary " A fire-department connection.

In many of our large cities the street-main supply is out of the question, because the available pressure is insufficient. While it is perfectly feasible to use a rotary or direct-acting fire-pump to supply the system, this would, in theatres where one fire-pump is already provided to supply the stand-pipe fire-valves, require a second pump of large capacity, for, as already stated, the fire-pump described heretofore is only to be used for the fire-valves and hose-streams, and to keep the fire-tank on the roof filled. It is therefore usual to choose for theatres the fifth combination, viz., to supply the sprinkler system primarily from a large roof tank, always kept full of water, and as an auxiliary supply to provide one or several outside fire-department connections, which enable the fire-department in case of a theatre fire to connect a steam fire-engine and to keep the sprinkler system on the stage supplied in case the roof tank has discharged its whole contents.

As regards the sprinkler-heads, any of the approved types, such as the Grinnell, Neracher, Walworth, Harkness, Hill, Kane, Buell, and others, may be used, if satisfactory to and approved by the building or fire department and the Board of Underwriters.

Sprinkler-heads must be made of non-corrodible metal, and the fusible solder must be of such a composition that it will be released at 155° to 160° Fahrenheit. The solder-joint of all sprinkler-heads must be so located that when it is acted upon by the fire and about to be released no water or its immediate chilling effect can reach the solder and prevent it from opening.

It is essential that the fusible solder-joint should be absolutely reliable, sensitive, and that it acts quickly and promptly.

All sprinklers must open full way, and each sprinkler outlet must deliver about one cubic foot of water per minute under the available pressure. Sprinkler-heads must be able to sustain a pressure of 300 lbs. per square inch without leaking, and they must be capable of operating under five pounds of water-pressure.

All sprinkler-heads are provided with distributors or deflectors, which divide the stream of water, as it strikes them, into a heavy shower, which thoroughly drenches the fire. These deflectors are stationary in some sprinklers, in others they are revolving or oscillating. The sprinkler-heads are placed either above or below the distributing pipes, the former position being slightly preferable because it secures perfect drainage and prevents the lodgment of rust, dirt, or sediment in the sprinklers.

The ordinary fusible solder-joint, which opens at 155 to 160° Fahrenheit, is not adapted for boiler rooms or drying-rooms, owing to the sometimes excessive heat in these rooms, and if automatic-sprinkler protection is desired in such places, a different alloy or solder having a higher degree of fusibility should be used.

The number of sprinklers necessary for an equipment depends upon the areas to be protected.

Sprinkler-heads must be set at no greater distance than 10 feet apart, and 5 feet away from walls or partitions, so that they protect an area of 10 feet in diameter. The rules of the New York Board of Fire Underwriters call for all portions of a building to be equipped by sprinklers, unless especially exempted because of being entirely fire-proof and containing only non-combustible materials, as for instance the loft over the auditorium; but in the stage portion of a theatre it is important that the space between ceiling or rigging-loft and stage roof be also protected. The largest number of sprinklers in a theatre is required to protect the rigging-loft and under it to protect the stage.

As a rule the piping of the sprinkler system consists of plain standard wrought-iron pipes, and if the pipes are constantly kept full of water there is probably not much danger of their rusting. In the dry-pipe system, however, it is essential that the supply-pipes should be rustproofed or galvanized, and where economy is no object, it is better to use such pipes also in the wet-pipe system.

As regards sizes for the sprinkler risers, distributing pipes, and branch-pipes to the individual sprinklers, the following rules should be strictly adhered to:

(*a*) The mains from the tank (or from the outside connections) must be of a capacity to supply at least 50 per cent of all sprinklers fed in the largest space enclosed by fire-walls, but where such a space would require less than 20 sprinklers the main should be capable of supplying 75 per cent of them.

(*b*) All branch-pipes supplying more than five and less than ten sprinklers shall be calculated to be of a capacity to supply 75 per cent of the sprinklers.

(*c*) Where over 30 sprinklers are supplied by one branch, it must have a capacity of supplying 50 per cent.

(*d*) No pipe smaller than $\frac{3}{4}$ inch shall be used, and only one sprinkler is to be fed from a $\frac{3}{4}$-inch pipe, provided the feed-pipe is not longer than 30 feet.

(*e*) The piping may be calculated from the following table, which gives the greatest number of sprinkler-heads to be fed from pipes of different diameter, those in column I being given by Mr. C. J. H. Woodbury, those in column II being the requirement of the New York Board of Fire Underwriters, those in column III being the Philadelphia Board of Underwriters' requirements for mercantile buildings and textile mills, and those in column IV for special risks, including theatres.

TABLE II.

Size of Pipe in Inches.	Number of Sprinkler-heads.			
	I.	II.	III.	IV. (theatres)
$\frac{3}{4}$	1	1	1	1
1	3	3	3	3
$1\frac{1}{4}$	6	5	6	5
$1\frac{1}{2}$	10	9	10	8
2	18	16	18	15
$2\frac{1}{2}$	28	25	28	24
3	46–48	36	46	35
$3\frac{1}{2}$	70–78	48	70	48
4	95–115	64	100	63
$4\frac{1}{2}$	......	82	...	...
5	......	100	250	100
6	......	144	over 250	150

All main branches and connections must be so arranged with drip-pipes and drip-valves that they can be completely emptied.

Each main and riser, and each distributing pipe on each floor, is to be provided with gate-valve, placed in the case of the mains near the tank. All valves to be full waterway gate-valves, with telltale indicators, and all valves must be strapped open by leather riveted or sealed straps.

All exposed sprinkler-heads must have guards

placed around them to protect them from damage, and this in theatres must be particularly observed in the case of sprinklers placed under the rigging-loft and the fly-galleries, with which sprinklers some of the hoisting machinery for the curtain or the borders and drops may come in contact. Nothing should be permitted to be hung from sprinkler-pipes, nor should the sprinkler-heads be painted, bronzed, or covered with whitewash.

The roof tank supplying the sprinkler system should be placed on the highest part of the stage roof, and must be elevated so that its bottom will be at least 12 feet above the level of the highest sprinkler. This location of the fire-tank on the roof is open to the objection that in case of fire the tank will fall with the collapse of the roof or the bearing walls and may thus become a source of danger to the firemen. Where it can be done it is preferable to build a separate tank tower.

The roof tank must be of large capacity, so as to be able to supply the sprinklers opened during a fire for a certain length of time.

Some require the capacity to be such as to fully operate for one hour at least 75 per cent of all sprinklers located in the largest room enclosed by fire-walls. For instance, if there are 50 sprinklers over the stage of a theatre, the capacity of the tank would figure out as follows: $50 \times \frac{3}{4} \times 1$ cubic foot $= 37.5$ cubic feet per minute or $37.5 \times 60 = 2250$ cubic feet per hour $= 16,875$ gallons tank capacity.

The New York Fire Underwriters' rules require the tank to be of a capacity to feed for fifteen minutes

50 per cent of all sprinklers supplied located in the largest compartment enclosed by fire-walls. This, for fifty sprinklers over stage, as above, would require $50 \times \frac{1}{2} \times 1 = 25$ cubic feet per minute = 375 cubic feet per $\frac{1}{4}$ hour, or a tank capacity of about 2812 gallons.

In the following is given an empirical table of the capacity and dimensions of round wooden tanks, for a given total number of sprinklers in a building.

TABLE III.

Number of Sprinklers.	Capacity of Tank in Gallons.	Length of Tank Staves.	Diameter of Tank.
100	2,500	7 feet	8 feet
150	3,500	9 "	9 "
200	5,000	10 "	10 "
300	6,000	10 "	11 "
500	7,500	10 "	12 "
800	10,000	12 "	14 "
1,000	15,000	12 "	16 "
1,200	20,000	15 "	16 "

The rules of the Philadelphia Board of Fire Underwriters require the capacity of the elevated tank to be as follows:

For 100 sprinklers or under, not less than 3000 gallons.

For 100 to 200 sprinklers or under, not less than 5000 gallons.

For 200 to 300 sprinklers, or under, not less than 7000 gallons.

For over 300 sprinklers, special sizes.

The Insurers' Automatic Fire Extinguisher Co. of New York gives the following table of relative proportions of size of pipe, capacity of tank, and number of sprinklers:

TABLE IV.

Size of Pipe.	Number of Sprinklers.	Size of Tank.
¾ inch	1	212 gallons
1 "	3	424 "
1¼ "	5	530 "
1½ "	9	959 "
2 "	14	1,696 "
2½ "	24	2,550 "
3 "	36	3,816 "
3½ "	49	5,194 "
4 "	64	6,784 "
5 "	100	10,000 "
6 "	144	10,000 "
7 "	196	10,000 "
8 "	256	10,000 "

At the Fifth Avenue Theatre in New York a roof tank of 6000 gallons capacity was installed for a total equipment of 227 sprinklers, distributed as follows: under the stage roof 44, under the gridiron or rigging-loft 50, under the first and second fly-galleries 12 and 9 respectively, under the stage 40, and for the dressing-rooms 72.

In the new Harrigan Theatre in New York the capacity of the roof tank was 4000 gallons and the total number of sprinklers installed was 105.

The roof tank should be painted and either frost-proofed or else provided with steam connection to prevent freezing of the water. It should have a 2-inch overflow and a 1½-inch emptying pipe carried to the roof. The pump riser, the tank main or sprinkler riser, and the tank-emptying pipe must all be securely protected with felting or other non-conducting material against freezing. The tank should also be provided with cover to keep the dust out and to retard evaporation of the water by exposure to the sun. The tank should also be fitted up with a substantially built outside ladder, permanently fastened to the tank, to facilitate inspections and cleaning. The cleaning of the tank should be attended to at least once a year. The tank falling main or sprinkler riser should enter at least 4 inches above the tank bottom to avoid drawing the sediment which settles at the tank bottom into the pipes. In winter time the steam circulation to the tank should be turned on.

The tank falling main must be provided with a check-valve and shut-off gate-valve. On the line of the falling main is placed the electric attachment operating the automatic fire-alarm. It consists of a special fitting connected by wiring with the automatic alarm-bells, and so arranged that when water starts to flow from the tank through this fitting to any sprinkler the fire-alarm gong rings automatically. This alarm can be readily tested at any time by opening up any drip-pipe on the sprinkler distribution system.

The tank is filled, as already described, from the theatre fire-pump, which is not to be otherwise connected with the sprinkler system, nor are the sprinkler risers to be used for attaching fire-valves. Neither must any water be drawn from the sprinkler tank for any other purpose.

The tank should be fitted up with a low-water alarm, and with a float indicator which always shows in the engine-room the height of water in the tank.

The watchman's alarm and the low-water alarm

should be tested often, and the outside alarm-gong must be suitably protected against the weather.

The fire-department connection to the sprinkler system, located in a conspicuous place on the outside of the building, must be of regulation fire-department size, and plainly marked. The pipe leading from it should be provided in the building with approved check-valve. The fire-department connection should be not less than 3 inches in diameter, with coupling and cap, and there should be one such pipe for every 50 sprinklers in the building, which may open at any one fire.

In case a fire-pump should be preferred to an outside connection, it should have a capacity of at least two fire-streams or 500 gallons per minute, and it should be arranged to operate automatically and to start instantly upon the opening of a sprinkler-head. If the pump draws from a suction-reservoir, its capacity should be sufficient to keep the pump supplied during at least one hour's full working capacity, which would make it equal to $500 \times 60 = 30,000$ gallons. The pump should be tested once a week, and should be located so as to be easy of access and free from the danger of fire.

In any case, wherever a wet-sprinkler system is installed, strict precautions should be taken against the pipes freezing, as in the event of a pipe bursting serious damage to the scenic decorations would result. As far as practicable, every sprinkler system should be tested periodically. This test should include some of the sprinkler-heads to make sure that they have not become inoperative from corrosion or sediment. The automatic alarm of the sprinkler system should also be regularly tested, and the batteries, the wiring, etc., inspected.

14. Perforated-Pipe System.

In some instances the stage of a theatre is fitted up with an open-sprinkler system, or else with a perforated-pipe system, in place of the automatic sprinklers. In both cases the pipe system is empty except when in action, the water, whether supplied from direct city pressure or from one or several elevated reservoirs, being held back by a valve, which is either operated by hand or else fastened by a hemp cord, which is supposed to burn at the outbreak of a fire, thereby releasing the valve. The perforated-pipe system is arranged near and above the rigging-loft, and consists of a number of sections or rows of pipes, each controlled by a separate valve or shut-off and fed from a large cross main. The valves are all operated from one point on the stage, and the attendant may put in operation all the lines simultaneously, or else only one or several lines separately, whereby the stage and stage-scenery is deluged at the place where the fire burns.

The open-sprinkler system is similar to, but perhaps more effective in action than, the perforated-pipe system. In some cases large sizes of revolving sprinkling nozzles are fitted up, and prove efficient fire-extinguishers, if one could but rely on the water being turned on promptly.

The stages in many French and German theatres are protected by perforated-pipe systems, whereas the English theatres have neither automatic sprinklers nor perforated pipes, because there being until quite recently no practical experience with the good results obtained by the former as in this country, these are considered unreliable, and also because the latter are looked upon as dangerous to property, although such an experienced fireman as Capt. Shaw of the London Fire Brigade concedes that "such arrangements would make the safety of an audience in case of fire absolutely and completely certain."

One of the Munich Royal Theatres was the first German theatre to be fitted up on a large scale with perforated sprinkling pipes, and in this theatre, as well as in the large new opera-house at Frankfort-on-the-Main, what threatened to become serious conflagrations have been put out by this apparatus. On Feb. 10th, 1881, in the third act of the opera "The Jewess," in the Frankfort-on-the-Main new opera-house, a gauze border caught fire from the border-lights. The iron curtain was at once lowered and one of the valves operating the open pipe-sprinkler system was opened, extinguishing the fire in a few minutes, so that after an intermission of only ten minutes the play could proceed.

The new Boston fire law requires "the proscenium-opening of every theatre to be provided with a $2\frac{1}{2}$-inch perforated iron pipe or equivalent arrangement of automatic or open sprinklers, so constructed as to form when in operation a complete water-curtain for the entire proscenium opening. There shall be for the rest of the stage a complete system of fire apparatus and perforated iron pipes, automatic or open sprinklers."

It seems unfortunate that this law should specify the

perforated pipe to be of iron. It is well known that plain wrought-iron pipe will soon rust if kept dry and exposed to the atmosphere, and the perforations in the pipe would soon stop up with rust, dust, or sediment, and become inoperative.

The only available material for perforated-pipe systems is brass or, better, copper pipe, and the systems installed in German theatres always consist of this material.

I do not know of any New York theatre fitted up with perforated pipes, whereas all the new theatres and many of the older ones have an automatic-sprinkler system. Where requirements of economy do not prevent it, I should favor the instalment, in addition to the automatic-sprinkler system, of a perforated pipe at the stage side of the proscenium-opening. This, when in action, would form an efficient water-curtain, and thus would be an additional protection of the fireproof iron or asbestos curtain, and thereby of the theatre audience in case of a stage fire.

15. FIRE-PAILS.

In addition to the automatic steam fire-pump and the automatic-sprinkler system every theatre should be provided with a large number of fire-pails or firebuckets. It is a well-known fact that more fires are put out annually by hand-buckets than by any other fire-extinguishing appliance. One bucket of water used at the proper moment, during the first minute of a fire just started, is of more value than the largest fire-pump half an hour later.

Fire-pails, therefore, constitute in theatres as in

other buildings the most useful and most effective of all interior non-automatic fire apparatus to subdue a local fire before the latter has gained any headway. Even in the hands of the most inexperienced employés, such fire-pails are a great protection and security.

A theatre requires a very ample supply of fire-buckets, judiciously distributed. The largest number should be placed on the stage, handy and ready for any emergency, but there should also be rows of pails in the fly-galleries, in the rigging-loft, in the corridors adjoining the actors' dressing-rooms, under the stage, in the engine-room, and likewise in the various divisions of the auditorium.

Fire-pails should not be kept on the floor, where they are liable to be overturned, but should be either hung on hooks or placed on shelves. They should be marked " Fire-pails " or " for fire only," and should be distinguished by being painted a bright-red color, to avoid their being used for other purposes and perhaps found missing or empty in case they are needed.

Fire-pails must always be kept full of water, which evaporates rapidly. The practice of putting covers on fire-pails to prevent evaporation is of doubtful expediency, as it renders the daily inspection of the fire-pails more difficult. There should be a penalty enforced for the use of the fire-pails for other purposes.

Fire-pails are made of many different materials. Besides the common wooden pail or bucket, there are galvanized-iron pails, paper pails and rubber pails, canvas, leather, indurated fibre, and even glass pails.

All fire-pails, of whatever material, should be durable

and strong, light, yet of good size, holding about eight to ten quarts of water. Wooden pails deteriorate rapidly, and if accidentally allowed to become dry, they warp and fall rapidly to pieces. Galvanized iron pails are better than wooden pails, but are subject to corrosion. Leather buckets, copper riveted and enamel painted on the inside, are very good but expensive. Canvas buckets are liable to leak, and are not used much. Neither the rubber nor the paper pails are very durable. In practical use the indurated-fibre pail has been found superior to all those named so far. It is usually made with a round bottom and set on special shelves with suitable holes. This form of pail has the advantage that the pail cannot be taken away and used for other purposes. The objection made against the round-bottomed pail that it cannot be set on the floor and used to supply hand force-pumps, is of no importance when the larger casks of water are provided. It is claimed for the indurated-fibre pail that it will hold a non-freezing solution (in the case of unheated rooms) which would corrode metal pails.

Three new forms of fire or "insurance" pails require special attention. The Harkness insurance pail is made of heavy, galvanized iron, of ten quarts capacity, and it is filled with a non-freezing fire-extinguishing liquid. To prevent its evaporation a glass cover is put over the liquid, and to protect the glass from breakage a perforated iron top is placed over the same, with a handle which is to be used to break the glass. The claim is made for this pail, that its contents do not freeze, corrode, evaporate, or deteriorate.

The Worcester chemical compartment fire-pail

(Lincoln's patent) is made of heavy galvanized iron, painted inside and outside. The chemicals are held in a dry powder form in a compartment in the cover, which is hermetically sealed at the top and bottom with heavy tin-foil. By breaking the tin-foil in the cover the chemical is precipitated into the water, forming a strong and fresh fire-extinguishing solution.

The chemicals of this fire-pail are therefore held in a dry state until required, and cannot deteriorate, which is claimed as the chief advantage of this type of pail.

Another chemical fire-pail, made in Worcester, Mass., under the Macomber patent, is a glass pail encased in a tin jacket. Glass was used, as it was feared that the chemical liquid might attack a galvanized-iron pail. This pail contains two gallons of a chemical non-freezing fire-extinguishing liquid, and is hermetically sealed by a soft tin-foil cover to prevent evaporation.

As to the number of fire-pails required for a theatre, it is usual in other buildings to estimate five pails for 5000 square feet or less of area, and to add one pail for each additional 1000 square feet; but for theatres this allowance, at least for the stage, is hardly sufficient. The New York Theatre Law requires four casks of water on the stage and at least two pails to each cask, making eight pails. There would, however, be no harm in doubling this allowance. The equipment of the Fifth Avenue Theatre comprised forty-eight galvanized-iron fire-pails and eighteen insurance pails.

16. Casks.

Casks or large barrels, containing about 40 or 50 gallons of water, should be kept in readiness on the stage, and it would be well to have some of these also in the under-stage, in the fly-galleries, and in the rigging-loft. There should be at least four casks on the stage, one in each fly-gallery on each side of the stage, and two in the rigging-loft. These casks should be kept filled with water at all times, and they are useful in case the fire-buckets are employed to extinguish a fire, and likewise as a source of supply for hand force-pumps.

For filling the casks by small hose, and also for refilling the fire-pails, fire stand-pipes are sometimes provided with hose bibbs located under each fire-valve and provided with detachable lever handles.

17. Portable Fire-extinguishing Apparatus.

In addition to the hand fire-buckets it is desirable to have on the stage and in the auditorium a number of portable fire-extinguishers. Sometimes common hand force-pumps are used, but much better than these are the fire-extinguishers working under pressure. Of these there are a great many types, which may be subdivided into chemical and pneumatic extinguishers, the former being loaded with chemicals generating carbonic-acid gas, while the latter are charged with compressed air.

The well-known Babcock and the Champion fire-extinguishers, both manufactured by the same firm, are examples of chemical fire-extinguishers, and throw a commingled stream of water and carbonic-acid gas. They are strongly made of heavy copper, in various sizes and capacities, ranging from a small portable hand apparatus to large chemical village engines.

The copper cylinder of the apparatus holds a solution of bicarbonate of sodium in water and also a sealed bottle of sulphuric acid. By turning a hand-wheel at the top of the extinguisher the bottle is broken, and immediately the two fluids mix, generating carbonic-acid gas.

The Harkness and the Miller fire-extinguishers are examples of the pneumatic extinguishers. In these the copper vessels hold a non-freezing fire-extinguishing liquid compound, and are charged with compressed air under a high pressure. By opening a lever cock the apparatus is put in operation, and no gas is generated until the liquid strikes the fire.

The advantage of the pneumatic extinguishers consist in their using no acids, which often do much damage when the apparatus is in operation. They are also safer and perhaps more easily handled than the chemical extinguishers.

All such apparatus should be located conveniently and within easy reach. The Fifth Avenue Theatre was fitted up with fifteen portable extinguishers located as follows: One on each side of the stage, four in the fly-galleries, two in the rigging-loft, two in the orchestra, two in the balcony, two in the gallery, and one in the space under the stage.

It is, of course, necessary that the chemical extinguishers be refilled as soon as possible after use, and it is also well to have the apparatus tested or inspected at intervals, as some of the working parts may have become rusted up.

Of hand-grenades as portable fire-extinguishers I will not speak, as they are inferior to both fire-pails and chemical or pneumatic extinguishers.

18. STEAM-JETS.

Although steam has not in practice been used to any great extent for fire-extinguishing purposes, it may prove in many cases very useful, particularly in concealed spaces or cramped corners not easily reached, as for instance up in the rigging-loft. It is feasible to provide in such places a few steam-nozzles, supplied from steam-pipes controlled by valves at the stage. The use of steam for extinguishing purposes on the stage, or in corridors or stairs, is a very doubtful measure, as the escaping steam would have a tendency to increase the confusion and panic during a fire.

19. SAND.

It is to be recommended to keep on the stage boxes of sand to be used for the extinguishment of oil fires, in case an oil-lamp, used in the representation of the play, should explode or catch fire on the stage.

20. FIRE-AXES AND FIRE-HOOKS.

Every theatre should be supplied with a liberal outfit of fire-axes with pick heads, polished axe-brackets, or fire-hooks on poles of various lengths to pull down burning scenery, and knives or hatchets in the fly-galleries and rigging-loft to cut down the ropes of hanging scenery.

The fire-hooks are, of course, to be placed only on the stage, and it is usual to provide one six-foot, one ten-foot, one fifteen-foot, and one twenty-foot fire-hook pole. The axes should be placed on polished boards at every point where a portable extinguisher is placed. In the Fifth Avenue Theatre there are to be found fifteen sets of these, distributed wherever fire danger is specially apprehended. The New York Building Law calls for " at least four axes to be placed on each tier on the stage and in the auditorium."

21. Minor Fire-extinguishing Devices.

It is, finally, necessary to keep in readiness on the stage during each performance some damp or slightly wet blankets to save actresses or ballet-girls whose dresses may have caught fire from the footlights or any other cause. Likewise should there be wet swabs or sponges attached to long poles to put out sparks from fireworks lodging in the scenery, particularly in the hanging borders. Asbestos sheets and asbestos gloves may also be provided for fire protection.

22. Life-saving Appliances.

In all theatres which are not strictly fire-proof there should be provided and kept in readiness in the theatre-manager's office, or somewhere else near the main entrance, a large jumping-net, to save people whose retreat has been cut off by fire or smoke. Rope and pulley fire-escapes at some of the upper windows, a few scaling-ladders, and a long extension-ladder would also prove valuable apparatus to have on hand in case of a theatre-fire catastrophe.

23. FIRE-ALARM AND WATCHMAN'S CLOCK.

In order to control the efficiency of the night-watchman it is advisable to fit up in each theatre a watchman's clock with electro-magnetic circuit.

Every theatre should also have a system of fire-alarm connected with the nearest engine-house of the city fire-department. The alarm system may consist either in a series of thermostats or heat detectors, set to operate at a certain degree of temperature, when an electric circuit is completed which rings the alarm-gong; or else in a series of small pipes, closed up with fusible solder, which open at a temperature of about 140° to 150° F., which pipes are filled with compressed air. The melting of the fusible plug causes the compressed air to escape, whereby a diaphragm is operated, giving the alarm.

24. WATER-SUPPLY FOR HOUSE SERVICE.

The foregoing description of the fire equipment of a theatre makes no mention of the water-service needed for the plumbing appliances in a theatre. Whatever is needed in this line should, in my judgment, be entirely separate from the fire-service.

The plumbing in some of our modern theatres is quite elaborate, consisting of a number of toilet-rooms for the use of the public; toilet-rooms for the theatre-manager, the stage employés, for the actors, actresses, and supers; lavatories in all the actors' dressing-rooms; house-tanks and house-pumps; hot-water tanks; etc.

It is, of course, necessary that all plumbing be arranged in a sanitary manner, and the general rules and principles of plumbing and house drainage, which the author has elaborately discussed elsewhere,* should be applied to theatres the same as to other classes of buildings.

---

* See the author's numerous works on Plumbing, House Drainage, Sewage Disposal, etc.

LITERATURE ON THEATRES.*

I. BOOKS.

ENGLISH.

BIRKMIRE, WM. H.—"The Planning and Construction of American Theatres." New York, 1896.

BUCKLE, JAMES GEO.—"Theatre Construction and Maintenance." London, 1888.

GERHARD, WM. PAUL.—"Theatre Fires and Panics: their Causes and Prevention." New York, 1896.

SACHS, EDWIN O., AND WOODROW, ERNEST A. E.—"Modern Opera-houses and Theatres." 3 vols. Vol. I. Published May, 1896. London.

SAUNDERS, GEO.—"Treatise on Theatres." 1790.

SHAW, CAPT. EYRE M.—"Fire in Theatres." Second edition. London, 1889.

GERMAN.

"Handbuch d. Architektur." Theil III, Bd. 6; Theil IV, Bd. 6. (Not yet published.)

"Baukunde des Architekten." Bd. I. 2 ; Bd. II.

BÜSING, PROF. F. W.—"Die Sicherheit in Theatern." Bd. 6. Heft 2 des Handbuchs d. Hygiene. Jena, 1894.

---

* The list given does not pretend to be complete, and the compiler will esteem it a favor if interested readers will call his attention to any omissions or inaccuracies.

CAVOS, A.—" Ueber die Einrichtung von Theater-Gebäuden." Leipsic, 1849.
 "Der Wieder-Aufbau d. Stadt-Theaters zu Riga." Riga, 1888.
DOEHRING, W.—" Handbuch des Feuer-Lösch-und-Rettungswesens." Berlin, 1881. Mit Atlas.
 "Das Feuer-Loeschwesen Berlins." Berlin, 1881.
FISCHER, HERM.—" Fortschritte der Architektur." Heft 5.
 "Heizung, Lüftung u. Beleuchtung der Theater." Darmstadt, 1894.
FLECK, PROF. D. H.—" Ueber Flammensicherheit u. Darstellung flammensicherer Gegenstände." Dresden, 1882.
FÖLSCH, AUG.—" Theater-Brände und die zur Verhütung derselben erforderlichen Schutzmassregeln." Hamburg, 1878.
 "Ergänzungsheft dazu." Hamburg, 1882.
 "Erinnerungen aus d. Leben eines Technikers.—Theater Brände." Hamburg, 1889.
GILARDONE, FRANZ.—" Handbuch des Theater-Lösch-und Rettungswesens." 3 vols. Strassburg and Hagenau, 1882, 1884.
 "Zum Brand der Komischen Oper in Paris." Hagenau, 1887.
 "Der Theater-Brand zu Exeter." Hagenau, 1888.
HASENAUER, CARL FREIHERR VON.—" Hofburg-Theater, das k. k., in Wien." 1889.
 "Hoftheater, das neue, zu Dresden."
HONIG, FRITZ.—" Loeschen und Retten." Köln, 1894.
HUDE, H. V. D., UND J. HENNICKE.—" Das Lessing-Theater in Berlin." 1889.
KLASEN, LUDWIG.—" Grundriss-Vorbilder von Gebäuden aller Art." Abth. VIII. Leipsic, 1884.
LANDHAUS.—" Ueber Akustik und Katakustik der Theater."
LANGHAUS, C. F.—"Das Stadt-Theater in Leipzig."
 " Das Victoria Theater in Berlin."

LINDNER, DR. GUSTAV.—"Das Feuer." Brünn, 1881.

LUCAE, R.—"Das Opernhaus zu Frankfurt a. M." Herausgegeben von Giesenberg und Becker. 1883.

MARCH, OTTO.—"Das Städt. Spiel- und Festhaus zu Worms." Berlin, 1890.

"Das Städtische Spiel- und Festhaus zu Worms." Vortrag, 1890.

NULL, VAN DER, UND SICCARDSBURG.—"Das k. k. Hof-Opernhaus in Wien."

"Das neue Opernhaus in Wien." 1879.

RUNGE, G.—"Das neue Opernhaus. (Academy of Music)," Philadelphia, 1882.

SCHINKEL, KARL FRIEDR.—"Dekorationen in den Königl. Hof-Theatern zu Berlin." 1874.

SEMPER, G.—"Das Königl. Hoftheater in Dresden." Brunswick, 1849.

"Das (alte) Königl. Hof-Theater in Dresden."

STAUDE, GUSTAV.—"Das Stadt-Theater zu Halle." 1886. Halle a. S.

STRACK, J. H.—"Das alt-griechische Theater-Gebäude." Potsdam, 1843.

STURMHÖFEL, A.—"Szene der Alten und Bühne der Neuzeit." Berlin, 1889.

STURMHOEFEL, A.—"Akustik des Baumeisters." Berlin, 1894.

TIETZ, E.—"Das Kroll'sche Etablissement in Berlin."

TITZ.—"Das Wallner-Theater und das Viktoria-Theater in Berlin."

VENERAND, W.—"Asbest und Feuerschutz." Wien, 1886.

WIESELER, F. J. A.—"Theater-Gebäude bei den Griechen und Römern." Göttingen, 1851.

FRENCH.

Chenevier, P.—" L'Incendie de l'Opéra-Comique de Paris, et le Théâtre de Sureté." Paris, 1888.

Contant, C.—" Parallèle des Principaux Théâtres Modernes de l'Europe et des Machines Théâtrales." Paris, 1820.

Daly et Davioud.—"Les Théâtres de la Place du Châtelet." Paris, 1874.

Garnier, Chas.—" Le Nouvel Opéra de Paris." Paris, 1875-81.

Garnier.—" Le Théâtre." Paris, 1871.

Gosset, A.—" Traité de la Construction des Théâtres." Paris, Baudry & Co., 1886.

Moynet, Georges.—" Trucs et Decors.—La Machinerie Théâtrale." Paris. (About 1892.)

Moynet, M. J.—" L'Envers du Théâtre." Paris, 1888.

Petit, Maxime.—" Les Grands Incendies." Paris, 1882.

Nuitter.—" Le Nouvel Opéra." Paris, Librairie Hachette, 1875.

Sauvageot, L., "Considerations sur la Construction des Théatres." Paris, 1877.

ITALIAN.

Donghi, Daniele.—" Sulla Sicurezza dei Teatri in Caso d'Incendio." Torino, 1888.

II. PAMPHLETS.

Gerhard, Wm. Paul.—" Theatre Fire Catastrophes and their Prevention." New York, 1894.

"The Essential Conditions of Safety in Theatres." New York, 1894.

"The Water-service and Fire Protection of Theatres." New York, 1894.

Hexamer, John C.—"On the Prevention of Fires in Theatres." Journal of Franklin Institute, August, 1882.

HEXAMER, JOHN C.—"The Construction and Interior Arrangement of Buildings designed to be used as Theatres." July, 1892.
"Causes of Fire." Journal of Franklin Institute April, July, and August, 1893.
HEXAMER, JOHN C., AND OTHERS.—"Report of Special Committee of the Franklin Institute on the Prevention of Fires in Theatres." June, 1883.
"Report of the Committee on Theatres and Public Halls of the Citizens' Association of Chicago." I., January, 1882. II., October, 1883. III., 1887.
SACHS, EDWIN O.—"Urban Fire Protection." London, 1895.
"Notes on Fire Brigades and Appliances of Continental Cities." London, 1894.
SHAW, EYRE M.—"Fire in Theatres." First edition. London, 1876.
YOUNG, A. H.—"Theatre Panics and their Cure." London : Batsford, 1896.

BECKER U. GIESENBERG.—"Das Opernhaus zu Frankfurt am Main." Frankf. a-M.
BOÖG, C., UND V. JONSTORFF, HANS FRH. V. JUPTNER.— "Zur Sicherheit des Lebens in den Theatern, mit besonderer Berücksichtigung der Theater-Brände." Wien, 1889.
FICHTNER, J.—"Die Feuer-Sicherheit im Theater." Striegau, 1882.
"Theater-Brände und deren Verhütung." Brünn, 1881.
FOCKT, C. TH.—"Der Brand des Ring-Theaters in Wien." Wien, 1881.
"Der Ring-Theater-Prozess in Wien." Wien, 1882.
FOELSCH, AUG.—"Ueber Theater-Brände, u. ueber die fü das neue Opernhaus in Wien getroffenen Sicherheits-Massregeln." Wien, 1870.

GILARDONE, FRANZ.—"Die neuesten Erfahrungen auf d. Gebiet der Theater Sicherheits-Frage." Hagenau, 1888.
"Die Theater-Brände des Jahres. 1888."
JUNG, LUDWIG.—"Für Feuerwehren." Viele Hefte.
"Die Feuersicherheit in oeffentlichen Gebäuden." München, 1879.
JUNK, D. V.—"Das Theater-System der Gegenwart u. Zukunft." Wien, 1884.
PATERA, AD.—"Ueber Flammen-Schutzmittel." Wien, 1871.
PROKOP, AUG.—"Die Sicherheit der Person im Theater, nebst einem Beitrag zur Theater-Bau-Frage." Brünn, 1882.
RICHTER, HEINR.—"Die Feuer-Sicherheit der Theater." Würzburg, 1886.
SCHOLLE, FRIEDR.—"Ueber Theater-Brände, deren Ursachen und Verhütung." Dresden, 1882.
"Ueber Imprägnations-Verfahren als Schutzmassregel gegen Feuersgefahr." Dresden.
STUDE.—"Ein Mahnwort an Jedermann ueber Feuer-Sicherheit u. Feuerschutz im Theater. Brünn, 1882.
ZIEMBINSKI, STANISLAW.—"Neue selbsthätige Feuer-Signal-Apparate, und ein Vorschlag z. Verhüten der Theater-Brände." Krakau, 1882.
"Dienstliche Anweisung für das Grossherz. Hoftheater zu Weimar." Weimar.
"Vorschläge des Nieder-Oesterreichischen Gewerbe-Vereins betreffend die Sicherung von Theatern gegen Feuers-Gefahr." Wien, 1882.
"Veröffentl. der Deutschen Edison Gesellschaft. I. Das Edison-Glühlicht, u. seine Bedeutung für das Rettungswesen." Berlin, 1883.
"Veröffentl. der Deutschen Edison Gesellschaft. II. Elektrische Beleuchtung von Theatern." Berlin, 1884.

"Vierteljahrsschrift der Wiener Freiw. Rettungsgesellschaft."
   January, March, 1883. Vol. II., 1, 2, 3 enth. Denkschrift ueber Paniken in Theatern.
"Projekt einer Theater-Reform der Gesellschaft zur Herstellung zeitgemässer Theater 'Asphaleia.'" Leipzig, 1882.

BROUARDEL, DR.—"La Mort dans les Théâtres," Revue Sanitaire de la Provence. No. 132–133, June, 1889.
CHENEVIER, P.—"La Question du Feu dans les Théâtres."
   "Extinction des Incendies des Théâtres."
   "La Sécurité des Spectateurs dans les Théâtres."
   "Reflexions sur l'Incendie de l'Opéra-Comique."
   "L'Opéra-Comique (aprés l'Incendie)."
   "La Défense des Théâtres Actuels Contre l'Incendie."
CHOQUET, DOCTEUR.—"Les Incendies dans les Théâtres." Paris, 1886.
FIGUIER.—"L'Incendie dans les Théâtres." 1881.
LÉGOYT, M.—"Statistique des Incendies dans les Théâtres." 1882.
PICCOLI, D. V.—"La Question du Feu dans les Théâtres." May, 1883.
TRIPIER, DR. A.—"Assainissement des Théâtres ; Ventilation, Éclairage, Chauffage." Paris, 1864.
   "Sur la Ventilation et l'Éclairage des Salles de Spectacle." Paris, 1858.
VIVIEN.—"Sureté dans les Théâtres."

III. ARTICLES IN JOURNALS AND MAGAZINES.

1. "The Construction and Arrangement of Theatres."—Report of a Committee appointed by the Austrian Society of Engineers and Architects. *Am. Architect*, 1882.

2. "The Prevention of Fires in Theatres."—Report of

the Special Committee of the Franklin Institute, 1883. *American Architect,* June 9, 1883.

3. " Prevention of Fires."—Report of the Committee of the Society of Arts, London, 1883. *American Architect,* June 16 and 23, 1883.

4. " Theatre Fires."—Four papers by " R. G.," in *Fire and-Water,* 1877.

5. " Theatre Fires and Remedies."—Two Papers by " J. A. F.," in *American Architect,* October 22 and November 12, 1887.

6. " Theatre Construction."—By E. A. E. Woodrow, Architect, Four parts, *Inland Architect,* August, September, and October, 1890.

7. " Prevention and Extinction of Fires in Theatres."—By E. A. E. Woodrow. *Journal Society of Arts,* London, April 18, 1884.

8. " Fire, its Prevention and Extinction in Theatres."—By Arthur W. C. Shean, 1884. *Journal Society of Arts,* London, April 18, 1884.

9. " Prevention and Extinction of Fires in Theatres."—By E. A. E. Woodrow, 1892. *Journal Society of Arts,* July 8, 1892. Reprinted in *Building,* August 20, 1892.

10. " Theatrical Architecture." I–VI.—By E. A. E. Woodrow. *American Architect,* October 10 to December 19, 1891.

11. " Theatre Building Regulations." I–VII.—By E. A. E. Woodrow. *American Architect,* April 16 to August, 1892.

12. "American Dramatic Theatres."—Five articles by John A. Fox, Architect, in *American Architect,* July to September, 1879.

13. " Scenic Illusions and Scenic Appliances."—Paper by Percy Fitzgerald. *Journal Society of Arts,* March 18, 1887.

14. " Theatres."—By E. A. E. Woodrow. *American*

Architect, begun April 14, 1884, to October, 1895. Twenty-two numbers.

15. "Theatres."—By E. A. Woodrow. Forty-seven articles, *Building News*, London, July 15, 1892, to December 28, 1894.

16. "The Stage."—London Paper, July to October, 1886, articles on "Hygiene of the Stage."

17. "The Sanitation of Theatres."—By Ernest Turner. Reprinted in *Building*, October 24, 1891.

18. "Hygiene of the Theatre."—By W. E. Roth. *American Architect*, October 22, 1887.

19. "The Planning and Construction of American Theatres."—By W. H. Birkmire. Twelve articles, October 13, 1894, to March 7, 1896, *Building*.

20. "The Essential Conditions of Safety in Theatres."—By Wm. Paul Gerhard. *American Architect*, June 23, July 7, July 14, and July 21, 1894.

21. "Some Suggestions as to Theatre Architecture."—By Edwin O. Sachs. The *Builder*, December 15, 1894.

22. "The Protection of Theatres from Fire."—By E. A. E. Woodrow. *Building News*, September 27, 1895.

23. "The Safety of Theatre Audiences."—By Edwin O. Sachs. *Building News*, September 27, 1895.

24. "Modern Theatre Stages."—By E. O. Sachs. *Engineering*, London, I–VIII, January 17 and 31, February 14 and 28, March 13, April 10 and 24, and May 8. (Not yet concluded.)

25. "Concert Halls and Assembly Rooms."—By E. A. E. Woodrow. I, *Building News*, October 11, 1895. (Not yet concluded.)

26. "Theatre Fire Catastrophes and their Prevention."—*Scientific American Supplement*, October 27 and November 3, 1894, also *Fire and Water*, 1894.

27. "Water-supply and Fire Prevention of Theatres."—*Fire and Water*, 1894.

28. "Some Recent Developments in Theatre Planning."—*Building News*, March 25, 1892.

29. "An Anniversary—the Ring Theatre Fire."—*American Architect*, December 23, 1882.

30. "Paramount Requirements of a Large Opera-house." By Dankmar Adler. *Inland Architect and Building News*, October 1887.

31. "Essential Features of a Large Opera-house."—By J. C. Cady. *Inland Architect and Building News*, October, 1887.

32. "Behind the Scenes of an Opera-house.—By Gustav Kobbé.

33. "Fire Protection."—By Edwin O. Sachs. *Journal Society of Arts*, December 7, 1894.

34. "Theatres and Fire-proof Construction."—By Walter Embden. *Journal Society of Arts*, January 27, 1888.

35. "Electrical Stage Effects."—By Theodore Waters. *Electric Power*, May, 1896.

36. "Theatre Building for American Cities."—By Dankmar Adler, Architect. Two articles, Vol. VII, Nos. 5 and 6, August and September, 1894. *Engineering Magazine*, 1894.

37. "New Theatres at Berlin and their Safety against Fire and Panic."—By Edwin O. Sachs. *American Architect*, May 19, 1894.

38. "Bad Air in Theatres."—By C. S. Montgomery. *Engineering Magazine*, Vol. VIII, No. 2, May, 1892.

39. "Theatre Ventilation."—By John P. Seddon. *Building News*, February 15, 1884.

40. "Fifth Avenue Theatre."—*Scientific American*, Architects' and Builders' Edition, March to July, 1893.

41. "Fifth Avenue Theatre."—*Engineering Record*, July and August, 1891.

42. "Theatre Machinery."—*Frank Leslie's Monthly Magazine*, February, 1895.

43. "Literature on Theatres."—Compiled by W. Paul Gerhard, C.E. *American Architect*, June 27, 1896.

IV. RULES AND REGULATIONS.

Polizei-Verordnung, betreffend die bauliche Anlage u. innere Einrichtung von Theatern, Circusgeb. u. öffentl. Versammlungs-Räumen. Berlin, 1889.

Nachtrag dazu. Berlin, 1891.

Bestimmungen ueber die Bau-Art der von der Staats-Bau-Verwaltung auszuf. Gebäude unter besonderer Berücksichtigung der Verkehrs-Sicherheit. Berlin, 1892.

London County Council. Metropolitan Building Acts. Regulations made by Council, February 9, 1892, with respect to theatres.

Letter relating to Precautions against Fire, by the Lord Chamberlain, to theatre-managers, February 15, 1864.

V. OFFICIAL REPORTS.

Report from the Select Committee on "Theatrical Licenses and Regulations," together with the Proceedings of the Committee, Minutes of Evidence, Appendix and Index. London, 1866 (Blue Book).

Report from the Committee of the House of Commons on "Theatres." London, 1887 (Blue Book).

Report from the Select Committee on "Theatres and Places of Entertainment," together with the Proceedings of the Committee, Minutes of Evidence, Appendix and Index. (House of Commons Blue Book, June 2, 1892.)

SHORT-TITLE CATALOGUE

OF THE

PUBLICATIONS

OF

JOHN WILEY & SONS,

NEW YORK.

LONDON: CHAPMAN & HALL, LIMITED.

ARRANGED UNDER SUBJECTS.

Descriptive circulars sent on application.
Books marked with an asterisk are sold at *net* prices only.
All books are bound in cloth unless otherwise stated.

AGRICULTURE.

CATTLE FEEDING—DISEASES OF ANIMALS—GARDENING, ETC.

Armsby's Manual of Cattle Feeding	12mo,	$1 75
Downing's Fruit and Fruit Trees	8vo,	5 00
Kemp's Landscape Gardening	12mo,	2 50
Stockbridge's Rocks and Soils	8vo,	2 50
Lloyd's Science of Agriculture	8vo,	4 00
Loudon's Gardening for Ladies. (Downing.)	12mo,	1 50
Steel's Treatise on the Diseases of the Ox	8vo,	6 00
" Treatise on the Diseases of the Dog	8vo,	3 50
Grotenfelt's The Principles of Modern Dairy Practice. (Woll.)	12mo,	2 00

ARCHITECTURE.

BUILDING—CARPENTRY—STAIRS, ETC.

Berg's Buildings and Structures of American Railroads	4to,	7 50
Birkmire's Architectural Iron and Steel	8vo,	3 50
" Skeleton Construction in Buildings	8vo,	3 00

Birkmire's Compound Riveted Girders..................8vo,	$2 00
" American Theatres—Planning and Construction.8vo,	3 00
Carpenter's Heating and Ventilating of Buildings..........8vo,	3 00
Freitag's Architectural Engineering..................8vo,	2 50
Kidder's Architect and Builder's Pocket-book.....Morocco flap,	4 00
Hatfield's American House Carpenter..................8vo,	5 00
" Transverse Strains..............................8vo,	5 00
Monckton's Stair Building—Wood, Iron, and Stone........4to,	4 00
Gerhard's Sanitary House Inspection...................16mo,	1 00
Downing and Wightwick's Hints to Architects............8vo,	2 00
" Cottages.....................................8vo,	2 50
Holly's Carpenter and Joiner............................18mo,	75
Worcester's Small Hospitals—Establishment and Maintenance, including Atkinson's Suggestions for Hospital Architecture..................................12mo,	1 25
The World's Columbian Exposition of 1893.............4to,	2 50

ARMY, NAVY, Etc.

MILITARY ENGINEERING—ORDNANCE—PORT CHARGES, ETC.

Cooke's Naval Ordnance 8vo,	$12 50
Metcalfe's Ordnance and Gunnery...........12mo, with Atlas,	5 00
Ingalls's Handbook of Problems in Direct Fire............8vo,	4 00
" Ballistic Tables..........................8vo,	1 50
Bucknill's Submarine Mines and Torpedoes..............8vo,	4 00
Todd and Whall's Practical Seamanship 8vo,	7 50
Mahan's Advanced Guard............................18mo,	1 50
" Permanent Fortifications. (Mercur.).8vo, half morocco,	7 50
Wheeler's Siege Operations...........................8vo,	2 00
Woodhull's Notes on Military Hygiene........12mo, morocco,	2 50
Dietz's Soldier's First Aid....................12mo, morocco,	1 25
Young's Simple Elements of Navigation..12mo, morocco flaps,	2 50
Reed's Signal Service................................	50
Phelps's Practical Marine Surveying....................8vo,	2 50
Very's Navies of the World...............8vo, half morocco,	3 50
Bourne's Screw Propellers............................4to,	5 00

Hunter's Port Charges................8vo, half morocco,	$13	00
*Dredge's Modern French Artillery........4to, half morocco,	20	00
" Record of the Transportation Exhibits Building, World's Columbian Exposition of 1893..4to, half morocco,	15	00
Mercur's Elements of the Art of War...................8vo,	4	00
" Attack of Fortified Places...................12mo,	2	00
Chase's Screw Propellers..............................8vo,	3	00
Winthrop's Abridgment of Military Law.............12mo,	2	50
De Brack's Cavalry Outpost Duties. (Carr.)....18mo, morocco,	2	00
Cronkhite's Gunnery for Non-com. Officers.....18mo, morocco,	2	00
Dyer's Light Artillery..............................12mo,	3	00
Sharpe's Subsisting Armies.........................18mo,	1	25
" " " ....18mo, morocco,	1	50
Powell's Army Officer's Examiner...................12mo,	4	00
Hoff's Naval Tactics................................8vo,	1	50
Bruff's Ordnance and Gunnery........................8vo,	6	00

ASSAYING.

SMELTING—ORE DRESSING—ALLOYS, ETC.

Furman's Practical Assaying...........................8vo,	3	00
Wilson's Cyanide Processes...........................12mo,	1	50
Fletcher's Quant. Assaying with the Blowpipe..12mo, morocco,	1	50
Ricketts's Assaying and Assay Schemes...............8vo,	3	00
*Mitchell's Practical Assaying. (Crookes.)............8vo,	10	00
Thurston's Alloys, Brasses, and Bronzes...............8vo,	2	50
Kunhardt's Ore Dressing..............................8vo,	1	50
O'Driscoll's Treatment of Gold Ores...................8vo,	2	00

ASTRONOMY.

PRACTICAL, THEORETICAL, AND DESCRIPTIVE.

Michie and Harlow's Practical Astronomy..............8vo,	3	00
White's Theoretical and Descriptive Astronomy.........12mo,	2	00
Doolittle's Practical Astronomy........................8vo,	4	00
Craig's Azimuth 4to,	3	50
Gore's Elements of Geodesy..........................8vo,	2	50

BOTANY.
GARDENING FOR LADIES, ETC.

Westermaier's General Botany. (Schneider.)............8vo,	$2	00
Thomé's Structural Botany.............................18mo,	2	25
Baldwin's Orchids of New England................. ......8vo,	1	50
Loudon's Gardening for Ladies. (Downing.)...........12mo,	1	50

BRIDGES, ROOFS, Etc.
CANTILEVER—HIGHWAY—SUSPENSION.

Boller's Highway Bridges................................8vo,	2	00
* " The Thames River Bridge................4to, paper,	5	00
Burr's Stresses in Bridges................................8vo,	3	50
Merriman & Jacoby's Text-book of Roofs and Bridges. Part I., Stresses...8vo,	2	50
Merriman & Jacoby's Text-book of Roofs and Bridges. Part II., Graphic Statics.............................8vo,	2	50
Merriman & Jacoby's Text-book of Roofs and Bridges. Part III., Bridge Design.............................8vo,	5	00
Merriman & Jacoby's Text-book of Roofs and Bridges. Part IV., Continuous, Draw, Cantilever, Suspension, and Arched Bridges........................(*In preparation*).		
Crehore's Mechanics of the Girder........................8vo,	5	00
Du Bois's Strains in Framed Structures....................4to,	10	00
Greene's Roof Trusses...................................8vo,	1	25
" Bridge Trusses.................................8vo,	2	50
" Arches in Wood, etc..........................8vo,	2	50
Waddell's Iron Highway Bridges........................8vo,	4	00
Wood's Construction of Bridges and Roofs...............8vo,	2	00
Foster's Wooden Trestle Bridges........................4to,	5	00
* Morison's The Memphis Bridge.................Oblong 4to,	10	00
Johnson's Modern Framed Structures....................4to,	10	00

CHEMISTRY.
QUALITATIVE—QUANTITATIVE—ORGANIC—INORGANIC, ETC.

Fresenius's Qualitative Chemical Analysis. (Johnson.).....8vo,	4	00
" Quantitative Chemical Analysis. (Allen.).......8vo,	6	00
" * " " " (Bolton.)......8vo,	1	50

Crafts's Qualitative Analysis. (Schaeffer.)..............12mo,	$1	50
Perkins's Qualitative Analysis................ ....12mo,	1	00
Thorpe's Quantitative Chemical Analysis................18mo,	1	50
Classen's Analysis by Electrolysis. (Herrick.)..............8vo,	3	00
Stockbridge's Rocks and Soils...........................8vo,	2	50
O'Brine's Laboratory Guide to Chemical Analysis..........8vo,	2	00
Mixter's Elementary Text-book of Chemistry............12mo,	1	50
Wulling's Inorganic Phar. and Med. Chemistry..........12mo,	2	00
Mandel's Bio-chemical Laboratory.....................12mo,	1	50
Austen's Notes for Chemical Students..................12mo,		
Schimpf's Volumetric Analysis........................12mo,	2	50
Hammarsten's Physiological Chemistry (Mandel.)..........8vo,	4	00
Miller's Chemical Physics............................8vo,	2	00
Pinner's Organic Chemistry. (Austen.)................12mo,	1	50
Kolbe's Inorganic Chemistry..........................12mo,	1	50
Ricketts and Russell's Notes on Inorganic Chemistry (Non-metallic).........................Oblong 8vo, morocco,		75
Drechsel's Chemical Reactions. (Merrill.)..............12mo,	1	25
Adriance's Laboratory Calculations.....................12mo,	1	25
Troilius's Chemistry of Iron..........................8vo,	2	00
Allen's Tables for Iron Analysis.........................8vo,		
Nichols's Water Supply (Chemical and Sanitary)..........8vo,	2	50
Mason's " " " " " 8vo,	5	00
Spencer's Sugar Manufacturer's Handbook. 12mo, morocco flaps,	2	00
Wiechmann's Sugar Analysis.........................8vo,	2	50
" Chemical Lecture Notes..................12mo,	3	00

DRAWING.

ELEMENTARY—GEOMETRICAL—TOPOGRAPHICAL.

Hill's Shades and Shadows and Perspective....(*In preparation*)		
Mahan's Industrial Drawing. (Thompson.).........2 vols., 8vo,	3	50
MacCord's Kinematics...............................8vo,	5	00
" Mechanical Drawing........................8vo,	4	00
" Descriptive Geometry.......................8vo,	3	00
Reed's Topographical Drawing. (II. A.).................4to,	5	00
Smith's Topographical Drawing. (Macmillan.)............8vo,	2	50
Warren's Free-hand Drawing 12mo,	1	00

Warren's Drafting Instruments..............................12mo,	$1	25
" Projection Drawing..............................12mo,	1	50
" Linear Perspective..............................12mo,	1	00
" Plane Problems..............................12mo,	1	25
" Primary Geometry..............................12mo,		75
" Descriptive Geometry..................2 vols., 8vo,	3	50
" Problems and Theorems..........................8vo,	2	50
" Machine Construction..............2 vols., 8vo,	7	50
" Stereotomy—Stone Cutting.......................8vo,	2	50
" Higher Linear Perspective 8vo,	3	50
" Shades and Shadows............................8vo,	3	00
Whelpley's Letter Engraving.........................12mo,	2	00

ELECTRICITY AND MAGNETISM.
ILLUMINATION—BATTERIES—PHYSICS.

* Dredge's Electric Illuminations....2 vols., 4to, half morocco,	25	00
" " " Vol. II..................4to,	7	50
Niaudet's Electric Batteries. (Fishback.)..............12mo,	2	50
Anthony and Brackett's Text-book of Physics............8vo,	4	00
Cosmic Law of Thermal Repulsion......................18mo,		75
Thurston's Stationary Steam Engines for Electric Lighting Purposes...12mo,	1	50
Michie's Wave Motion Relating to Sound and Light,.......8vo,	4	00
Barker's Deep-sea Soundings............................8vo,	2	00
Holman's Precision of Measurements....................8vo,	2	00
Tillman's Heat...8vo,	1	50
Gilbert's De-magnete. (Mottelay.)......................8vo,	2	50
Benjamin's Voltaic Cell.................................8vo,	3	00
Reagan's Steam and Electrical Locomotives.............12mo	2	00

ENGINEERING.
CIVIL—MECHANICAL—SANITARY, ETC.

* Trautwine's Cross-section...........................Sheet,		25
* " Civil Engineer's Pocket-book...12mo, mor. flaps,	5	00
* " Excavations and Embankments.............8vo,	2	00
* " Laying Out Curves...........12mo, morocco,	2	50
Hudson's Excavation Tables. Vol. II.................. 8vo,	1	00

Searles's Field Engineering............12mo, morocco flaps,	$3 00
" Railroad Spiral................12mo, morocco flaps,	1 50
Godwin's Railroad Engineer's Field-book.12mo, pocket-bk. form,	2 50
Butts's Engineer's Field-book................12mo, morocco,	2 50
Gore's Elements of Goodesy.........................8vo,	2 50
Wellington's Location of Railways..................8vo,	5 00
*Dredge's Penn. Railroad Construction, etc... Folio, half mor.,	20 00
Smith's Cable Tramways.........................4to,	2 50
" Wire Manufacture and Uses....................4to,	.3 00
Mahan's Civil Engineering. (Wood.)..................8vo,	5 00
Wheeler's Civil Engineering.......................8vo,	4 00
Mosely's Mechanical Engineering. (Mahan.).............8vo,	5 00
Johnson's Theory and Practice of Surveying............8vo,	4 00
" Stadia Reduction Diagram..Sheet, 22¼ × 28¼ inches,	50
*Drinker's Tunnelling..................4to, half morocco,	25 00
Eissler's Explosives—Nitroglycerine and Dynamite........8vo,	4 00
Foster's Wooden Trestle Bridges.....................4to,	5 00
Ruffner's Non-tidal Rivers.........................8vo,	1 25
Greene's Roof Trusses 8vo,	1 25
" Bridge Trusses..............................8vo,	2 50
" Arches in Wood, etc........................8vo,	2 50
Church's Mechanics of Engineering—Solids and Fluids....8vo,	6 00
" Notes and Examples in Mechanics..............8vo,	2 00
Howe's Retaining Walls (New Edition.)...............12mo,	1 25
Wegmann's Construction of Masonry Dams..............4to,	5 00
Thurston's Materials of Construction..................8vo,	5 00
Baker's Masonry Construction.......................8vo,	5 00
" Surveying Instruments.....................12mo,	3 00
Warren's Stereotomy—Stone Cutting..................8vo,	2 50
Nichols's Water Supply (Chemical and Sanitary).........8vo,	2 50
Mason's " " " " " 8vo,	5 00
Gerhard's Sanitary House Inspection.................16mo,	1 00
Kirkwood's Lead Pipe for Service Pipe................8vo,	1 50
Wolff's Windmill as a Prime Mover..................8vo,	3 00
Howard's Transition Curve Field-book.....12mo, morocco flap,	1 50
Crandall's The Transition Curve.............12mo, morocco,	1 50

Crandall's Earthwork Tables 8vo,	$1	50
Patton's Civil Engineering..........................8vo,	7	50
" Foundations..........................8vo,	5	00
Carpenter's Experimental Engineering 8vo,	6	00
Webb's Engineering Instruments.............12mo, morocco,	1	00
Black's U. S. Public Works..........................4to,	5	00
Merriman and Brook's Handbook for Surveyors....12mo, mor.,	2	00
Merriman's Retaining Walls and Masonry Dams..........8vo,	2	00
" Geodetic Surveying......................8vo,	2	00
Kiersted's Sewage Disposal..... ....12mo,	1	25
Siebert and Biggin's Modern Stone Cutting and Masonry...8vo,	1	50
Kent's Mechanical Engineer's Pocket-book.....12mo, morocco,	5	00

HYDRAULICS.

WATER-WHEELS—WINDMILLS—SERVICE PIPE—DRAINAGE, ETC.

Weisbach's Hydraulics. (Du Bois.).....................8vo,	5	00
Merriman's Treatise on Hydraulics.....................8vo,	4	00
Ganguillet & Kutter's Flow of Water. (Hering & Trautwine.).8vo,	4	00
Nichols's Water Supply (Chemical and Sanitary)..........8vo,	2	50
Wolff's Windmill as a Prime Mover......................8vo,	3	00
Ferrel's Treatise on the Winds, Cyclones, and Tornadoes...8vo,	4	00
Kirkwood's Lead Pipe for Service Pipe 8vo,	1	50
Ruffner's Improvement for Non-tidal Rivers..............8vo,	1	25
Wilson's Irrigation Engineering.... ....8vo,	4	00
Bovey's Treatise on Hydraulics..........................8vo,	4	00
Wegmann's Water Supply of the City of New York.......4to,	10	00
Hazen's Filtration of Public Water Supply................8vo,	2	00
Mason's Water Supply—Chemical and Sanitary............8vo,	5	00
Wood's Theory of Turbines.... ....8vo,	2	50

MANUFACTURES.

ANILINE—BOILERS—EXPLOSIVES—IRON—SUGAR—WATCHES— WOOLLENS, ETC.

Metcalfe's Cost of Manufactures.........................8vo,	5	00
Metcalf's Steel (Manual for Steel Users).................12mo,	2	00
Allen's Tables for Iron Analysis..........................8vo,		

West's American Foundry Practice................12mo,	$2	50
" Moulder's Text-book 12mo,	2	50
Spencer's Sugar Manufacturer's Handbook....12mo, mor. flap,	2	00
Wiechmann's Sugar Analysis...........................8vo,	2	50
Beaumont's Woollen and Worsted Manufacture.........12mo,	1	50
*Reisig's Guide to Piece Dyeing........................8vo,	25	00
Eissler's Explosives, Nitroglycerine and Dynamite........8vo,	4	00
Reimann's Aniline Colors. (Crookes.)................8vo,	2	50
Ford's Boiler Making for Boiler Makers................18mo,	1	00
Thurston's Manual of Steam Boilers....................8vo,	5	00
Booth's Clock and Watch Maker's Manual...............12mo,	2	00
Holly's Saw Filing......................................18mo,		75
Svedelius's Handbook for Charcoal Burners.............12mo,	1	50
The Lathe and Its Uses................................8vo,	6	00
Woodbury's Fire Protection of Mills....................8vo,	2	50
Bolland's The Iron Founder...........................12mo,	2	50
" " " " Supplement................12mo,	2	50
" Encyclopædia of Founding Terms............12mo,	3	00
Bouvier's Handbook on Oil Painting....................12mo,	2	00
Steven's House Painting...............................18mo,		75

MATERIALS OF ENGINEERING.

STRENGTH—ELASTICITY—RESISTANCE, ETC.

Thurston's Materials of Engineering..............3 vols., 8vo,	8	00
Vol. I., Non-metallic..............................8vo,	2	00
Vol. II., Iron and Steel............................8vo,	3	50
Vol. III., Alloys, Brasses, and Bronzes.............8vo,	2	50
Thurston's Materials of Construction....................8vo,	5	00
Baker's Masonry Construction..........................8vo,	5	00
Lanza's Applied Mechanics............................8vo,	7	50
" Strength of Wooden Columns.............8vo, paper,		50
Wood's Resistance of Materials........................8vo,	2	00
Weyrauch's Strength of Iron and Steel. (Du Bois.).......8vo,	1	50
Burr's Elasticity and Resistance of Materials..............8vo,	5	00
Merriman's Mechanics of Materials.....................8vo,	4	00
Church's Mechanic's of Engineering—Solids and Fluids....8vo,	6	00

Beardslee and Kent's Strength of Wrought Iron..........8vo,	$1	50
Hatfield's Transverse Strains............................8vo,	5	00
Du Bois's Strains in Framed Structures..................4to,	10	00
Merrill's Stones for Building and Decoration.............8vo,	5	00
Bovey's Strength of Materials...........................8vo,	7	50
Spalding's Roads and Pavements........................12mo,	2	00
Rockwell's Roads and Pavements in France.............12mo,	1	25
Byrne's Highway Construction...........................8vo,	5	00
Patton's Treatise on Foundations........................8vo,	5	00

MATHEMATICS.

CALCULUS—GEOMETRY—TRIGONOMETRY, ETC.

Rice and Johnson's Differential Calculus..................8vo,	3	50
" Abridgment of Differential Calculus....8vo,	1	50
" Differential and Integral Calculus, 2 vols. in 1, 12mo,	2	50
Johnson's Integral Calculus............................12mo,	1	50
" Curve Tracing...............................12mo,	1	00
" Differential Equations—Ordinary and Partial.....8vo,	3	50
" Least Squares...............................12mo,	1	50
Craig's Linear Differential Equations.....................8vo,	5	00
Merriman and Woodward's Higher Mathematics...........8vo,		
Bass's Differential Calculus............................12mo,		
Halsted's Synthetic Geometry...........................8vo,	1	50
" Elements of Geometry.........................8vo,	1	75
Chapman's Theory of Equations........................12mo,	1	50
Merriman's Method of Least Squares....................8vo,	2	00
Compton's Logarithmic Computations...................12mo,	1	50
Davis's Introduction to the Logic of Algebra..............8vo,	1	50
Warren's Primary Geometry...........................12mo,		75
" Plane Problems............................. 12mo,	1	25
" Descriptive Geometry..................2 vols., 8vo,	3	50
" Problems and Theorems.......................8vo,	2	50
" Higher Linear Perspective.....................8vo,	3	50
" Free-hand Drawing..........................12mo,	1	00
" Drafting Instruments........................12mo,	1	25

Warren's Projection Drawing................................12mo,	$1	50
" Linear Perspective...........................12mo,	1	00
" Plane Problems...............................12mo,	1	25
Searles's Elements of Geometry. 8vo,	1	50
Brigg's Plane Analytical Geometry..........................12mo,	1	00
Wood's Co-ordinate Geometry............................8vo,	2	00
" Trigonometry.....................................12mo,	1	00
Mahan's Descriptive Geometry (Stone Cutting)..............8vo,	1	50
Woolf's Descriptive Geometry......................Royal 8vo,	3	00
Ludlow's Trigonometry with Tables. (Bass.)..............8vo,	3	00
" Logarithmic and Other Tables. (Bass.).........8vo,	2	00
Baker's Elliptic Functions...............................8vo,	1	50
Parker's Quadrature of the Circle.......................8vo,	2	50
Totten's Metrology...8vo,	2	50
Ballard's Pyramid Problem 8vo,	1	50
Barnard's Pyramid Problem.............................8vo,	1	50

MECHANICS—MACHINERY.

TEXT-BOOKS AND PRACTICAL WORKS.

Dana's Elementary Mechanics........................12mo,	1	50
Wood's " " 12mo,	1	25
" " " Supplement and Key............	1	25
" Analytical Mechanics............................8vo,	3	00
Michie's Analytical Mechanics............................8vo,	4	00
Merriman's Mechanics of Materials......................8vo,	4	00
Church's Mechanics of Engineering......................8vo,	6	00
" Notes and Examples in Mechanics..............8vo,	2	00
Mosely's Mechanical Engineering. (Mahan.)..............8vo,	5	00
Weisbach's Mechanics of Engineering. Vol. III., Part I., Sec. I. (Klein.)................................8vo,	5	00
Weisbach's Mechanics of Engineering. Vol. III., Part I., Sec. II. (Klein.)..............................8vo,	5	00
Weisbach's Hydraulics and Hydraulic Motors. (Du Bois.)..8vo,	5	00
" Steam Engines. (Du Bois.)..................8vo,	5	00
Lanza's Applied Mechanics..............................8vo,	7	50

Crehore's Mechanics of the Girder......8vo,	$5 00
MacCord's Kinematics......8vo,	5 00
Thurston's Friction and Lost Work......8vo,	3 00
" The Animal as a Machine......12mo,	1 00
Hall's Car Lubrication......12mo,	1 00
Warren's Machine Construction......2 vols., 8vo,	7 50
Chordal's Letters to Mechanics......12mo,	2 00
The Lathe and Its Uses......8vo,	6 00
Cromwell's Toothed Gearing......12mo,	1 50
" Belts and Pulleys......12mo,	1 50
Du Bois's Mechanics. Vol. I., Kinematics......8vo,	3 50
" " Vol. II., Statics......8vo,	4 00
" " Vol. III., Kinetics......8vo,	3 50
Dredge's Trans. Exhibits Building, World Exposition, 4to, half morocco,	15 00
Flather's Dynamometers......12mo,	2 00
" Rope Driving......12mo,	2 00
Richards's Compressed Air......12mo,	1 50
Smith's Press-working of Metals......8vo,	3 00
Holly's Saw Filing......18mo,	75
Fitzgerald's Boston Machinist......18mo,	1 00
Baldwin's Steam Heating for Buildings......12mo,	2 50
Metcalfe's Cost of Manufactures......8vo,	5 00
Benjamin's Wrinkles and Recipes......12mo,	2 00
Dingey's Machinery Pattern Making......12mo,	2 00

METALLURGY.

IRON—GOLD—SILVER—ALLOYS, ETC.

Egleston's Metallurgy of Silver......8vo,	7 50
" Gold and Mercury......8vo,	7 50
" Weights and Measures, Tables......18mo,	75
" Catalogue of Minerals......8vo,	2 50
O'Driscoll's Treatment of Gold Ores......8vo,	2 00
* Kerl's Metallurgy—Copper and Iron......8vo,	15 00
* " " Steel, Fuel, etc......8vo,	15 00

Thurston's Iron and Steel.....................8vo,	$3	50
" Alloys........................8vo,	2	50
Troilius's Chemistry of Iron......................8vo,	2	00
Kunhardt's Ore Dressing in Europe......................8vo,	1	50
Weyrauch's Strength of Iron and Steel. (Du Bois.)........8vo,	1	50
Beardslee and Kent's Strength of Wrought Iron..........8vo,	1	50
Compton's First Lessons in Metal Working.............12mo,	1	50
West's American Foundry Practice....................12mo,	2	50
" Moulder's Text-book........................12mo,	2	50

MINERALOGY AND MINING.

MINE ACCIDENTS—VENTILATION—ORE DRESSING, ETC.

Dana's Descriptive Mineralogy. (E. S.).....8vo, half morocco,	12	50
" Mineralogy and Petrography. (J. D.)..........12mo,	2	00
" Text-book of Mineralogy. (E. S.)................8vo,	3	50
" Minerals and How to Study Them. (E. S.).......12mo,	1	50
" American Localities of Minerals..................8vo,	1	00
Brush and Dana's Determinative Mineralogy.. ....8vo,	3	50
Rosenbusch's Microscopical Physiography of Minerals and Rocks. (Iddings.)..............................8vo,	5	00
Hussak's Rock-forming Minerals. (Smith.)..............8vo,	2	00
Williams's Lithology......................................8vo,	3	00
Chester's Catalogue of Minerals..........................8vo,	1	25
" Dictionary of the Names of Minerals............8vo,	3	00
Egleston's Catalogue of Minerals and Synonyms..........8vo,	2	50
Goodyear's Coal Mines of the Western Coast...........12mo,	2	50
Kunhardt's Ore Dressing in Europe......................8vo,	1	50
Sawyer's Accidents in Mines........................8vo,	7	00
Wilson's Mine Ventilation...........................16mo,	1	25
Boyd's Resources of South Western Virginia.............8vo,	3	00
" Map of South Western Virginia.....Pocket-book form,	2	00
Stockbridge's Rocks and Soils...........................8vo,	2	50
Eissler's Explosives—Nitroglycerine and Dynamite........8vo,	4	00

*Drinker's Tunnelling, Explosives, Compounds, and Rock Drills.
4to, half morocco, $25 00
Beard's Ventilation of Mines..................12mo, 2 50
Ihlseng's Manual of Mining............................8vo, 4 00

STEAM AND ELECTRICAL ENGINES, BOILERS, Etc.
STATIONARY—MARINE—LOCOMOTIVE—GAS ENGINES, ETC.

Weisbach's Steam Engine. (Du Bois.)..................8vo, 5 00
Thurston's Engine and Boiler Trials......................8vo, 5 00
" Philosophy of the Steam Engine............12mo, 75
" Stationary Steam Engines..................12mo, 1 50
" Boiler Explosion.......................12mo, 1 50
" Steam-boiler Construction and Operation.......8vo,
" Reflection on the Motive Power of Heat. (Carnot.)
12mo, 2 00
Thurston's Manual of the Steam Engine. Part I., Structure
and Theory..8vo, 7 50
Thurston's Manual of the Steam Engine. Part II., Design,
Construction, and Operation......................8vo, 7 50
2 parts, 12 00
Röntgen's Thermodynamics. (Du Bois.)................8vo, 5 00
Peabody's Thermodynamics of the Steam Engine......... 8vo, 5 00
" Valve Gears for the Steam-Engine.............8vo, 2 50
" Tables of Saturated Steam....................8vo, 1 00
Wood's Thermodynamics, Heat Motors, etc..............8vo, 4 00
Pupin and Osterberg's Thermodynamics.................12mo, 1 25
Kneass's Practice and Theory of the Injector 8vo, 1 50
Reagan's Steam and Electrical Locomotives.............12mo, 2 00
Meyer's Modern Locomotive Construction.................4to, 10 00
Whitham's Steam-engine Design 8vo, 6 00
" Constructive Steam Engineering..............8vo, 10 00
Hemenway's Indicator Practice......................12mo, 2 00
Pray's Twenty Years with the Indicator...........Royal 8vo, 2 50
Spangler's Valve Gears................................8vo, 2 50
*Maw's Marine Engines................Folio, half morocco, 18 00
Trowbridge's Stationary Steam Engines ....4to, boards, 2 50

Ford's Boiler Making for Boiler Makers	18mo,	$1 00
Wilson's Steam Boilers. (Flather.)	12mo,	2 50
Baldwin's Steam Heating for Buildings	12mo,	2 50
Hoadley's Warm-blast Furnace	8vo,	1 50
Sinclair's Locomotive Running	12mo,	2 00
Clerk's Gas Engine	12mo,	

TABLES, WEIGHTS, AND MEASURES.

FOR ENGINEERS, MECHANICS, ACTUARIES—METRIC TABLES, ETC.

Crandall's Railway and Earthwork Tables	8vo,	1 50
Johnson's Stadia and Earthwork Tables	8vo,	1 25
Bixby's Graphical Computing Tables	Sheet,	25
Compton's Logarithms	12mo,	1 50
Ludlow's Logarithmic and Other Tables. (Bass.)	12mo,	2 00
Thurston's Conversion Tables	8vo,	1 00
Egleston's Weights and Measures	18mo,	75
Totten's Metrology	8vo,	2 50
Fisher's Table of Cubic Yards	Cardboard,	25
Hudson's Excavation Tables. Vol. II	8vo,	1 00

VENTILATION.

STEAM HEATING—HOUSE INSPECTION—MINE VENTILATION.

Beard's Ventilation of Mines	12mo,	2 50
Baldwin's Steam Heating	12mo,	2 50
Reid's Ventilation of American Dwellings	12mo,	1 50
Mott's The Air We Breathe, and Ventilation	16mo,	1 00
Gerhard's Sanitary House Inspection	Square 16mo,	1 00
Wilson's Mine Ventilation	16mo,	1 25
Carpenter's Heating and Ventilating of Buildings	8vo,	3 00

MISCELLANEOUS PUBLICATIONS.

Alcott's Gems, Sentiment, Language	Gilt edges,	5 00
Bailey's The New Tale of a Tub	8vo,	75
Ballard's Solution of the Pyramid Problem	8vo,	1 50
Barnard's The Metrological System of the Great Pyramid	8vo,	1 50

* Wiley's Yosemite, Alaska, and Yellowstone 4to,	$3 00
Emmon's Geological Guide-book of the Rocky Mountains..8vo,	1 50
Ferrel's Treatise on the Winds..........................8vo,	4 00
Perkins's Cornell University.....................Oblong 4to,	1 50
Ricketts's History of Rensselaer Polytechnic Institute.....8vo,	3 00
Mott's The Fallacy of the Present Theory of Sound..Sq. 16mo,	1 00
Rotherham's The New Testament Critically Emphathized. 12mo,	1 50
Totten's An Important Question in Metrology............8vo,	2 50
Whitehouse's Lake Mœris............................Paper,	25

HEBREW AND CHALDEE TEXT-BOOKS.

For Schools and Theological Seminaries.

Gesenius's Hebrew and Chaldee Lexicon to Old Testament. (Tregelles.)....................Small 4to, half morocco,	5 00
Green's Grammar of the Hebrew Language (New Edition).8vo,	3 00
" Elementary Hebrew Grammar................12mo,	1 25
" Hebrew Chrestomathy.........................8vo,	2 00
Letteris's Hebrew Bible (Massoretic Notes in English). 8vo, arabesque,	2 25
Luzzato's Grammar of the Biblical Chaldaic Language and the Talmud Babli Idioms...........................12mo,	1 50

MEDICAL.

Bull's Maternal Management in Health and Disease.......12mo,	1 00
Mott's Composition, Digestibility, and Nutritive Value of Food. Large mounted chart,	1 25
Steel's Treatise on the Diseases of the Ox.... ....8vo,	6 00
" Treatise on the Diseases of the Dog...............8vo,	3 50
Worcester's Small Hospitals—Establishment and Maintenance, including Atkinson's Suggestions for Hospital Architecture......................................12mo,	1 25
Hammarsten's Physiological Chemistry. (Mandel.)........8vo,	4 00

www.ingramcontent.com/pod-product-compliance
Lightning Source LLC
Chambersburg PA
CBHW020927230426
43666CB00008B/1594